Building Technic
VOLUME ONE: STRUCTURE

Building Techniques

HAROLD KING, M.A., A.R.I.B.A.
*Senior Lecturer for Building Technology,
School of Architecture,
University of Newcastle upon Tyne*

DENZIL NIELD, A.R.I.B.A.

VOLUME ONE: STRUCTURE
S.I. Edition

LONDON
CHAPMAN AND HALL

A Halsted Press Book
John Wiley & Sons, New York

First published as *Building Construction Illustrated* 1952
by E. and F. N. Spon Ltd., 11 New Fetter Lane, London EC4P 4EE
Revised edition in two volumes
Under the present title published 1967
First published in Science Paperbacks 1967
Reprinted 1972 and 1974
Third edition 1976
© 1967, 1976 Harold King and Denzil Nield

Typeset by Santype International Ltd.
(Coldtype Division), Salisbury, Wiltshire
Printed in Great Britain by Billing & Sons Ltd.,
Guildford, London and Worcester

ISBN 0 412 21330 3

This paperback edition is
sold subject to the condition that it
shall not, by way of trade or otherwise, be
lent, re-sold, hired out, or otherwise circulated
without the publisher's prior consent in any form of
binding or cover other than that in which it is
published and without a similar condition
including this condition being imposed
on the subsequent purchaser

All rights reserved. No part of
this book may be reprinted, or reproduced
or utilized in any form or by any electronic,
mechanical or other means, now known or hereafter
invented, including photocopying and recording,
or in any information storage or retrieval
system, without permission in writing
from the publisher

Distributed in the U.S.A. by Halsted Press,
a Division of John Wiley & Sons, Inc., New York

Contents

PREFACE	*page*	xi
AUTHOR'S NOTE TO FIRST EDITION		xiii
AUTHOR'S NOTE TO REVISED (S.I.) EDITION		xv
THE METRIC SYSTEM		xvii

1	THE SITE	1
	Choice of site	1
	Site survey	1
	Setting out	3
	Removal of top soil	5
	Site drainage	5
	Excavation	5
	Safety in excavations	6
2	EXTERNAL WORKS	8
	Estate roads	8
	Road drainage	10
	Edging	11
	Paved surfaces	13
	Kerbs and trim	16
	Fences	17
	Gates	20
	Boundary walls	20
	Copings	24
3	LOAD-BEARING CONSTRUCTION	26
	Walling	26
	Insulation	27
	Foundations	30

Settlement	31
Types of foundations	31
Characteristics of sub-soils	31
Safe bearing capacity	33
Wide strip concrete foundations	35
Deep strip concrete foundations	36
Short concrete pile foundations	37
Concrete raft	38
'Made-up' ground	38
Reinforced concrete foundations	38
Concrete in foundations	39
Brickwork materials	42
Classification of bricks	43
Mortar	45
Mortar Mix	46
Mortar with plasticizer	46
Laying bricks	47
Jointing and pointing	47
Brick walls	48
Cross-walling	50
Arches and lintels	52
Damp-proof courses	56
Damp-proof membranes	57
Rendering	57
Cavity wall construction	58
Fireplaces and flues	61
Flue for domestic boiler	65
Stonemasonry	66

4 FRAMED CONSTRUCTION
Structural steelwork	68
Timber framing	74

5 FLOORS
Timber	77
Timber upper floors	80
Stress grading	81
Reinforced concrete upper floors	85
Formwork	85

CONTENTS · vii

Pre-stressed concrete beams	87
Ground floors in timber	88
Site concrete	89
Under-floor ventilation	90
Floorboarding	90
Solid ground floors	93
Condensation	93

6 ROOFS — 94

Structure	94
Insulation	95
Fire resistance	95
Wind-loading on roofs	96
Snow-loading	96
Flat roofs	96
Asphalt	97
Built-up felt	97
Vented underlays	97
Promenade roofing	98
Vapour barrier	98
Metal sheet roofing	99
Compound sheet roofing	100
Thermo-plastic sheet roofing	100
Inverted roof	102
Concrete flat roofs	102
Timber flat roofs	104
Lead-covered flat roofs	106
Pitched roofs	108
Materials for coverings	108
Plain tiles	108
Single-lap tiles	110
Interlocking tiles	110
Slates	111
Systems of pitched roof construction	112
Single roof systems	113
Purlin roof systems	114
Trussed rafter	115
Roof truss systems	116
Details of purlin roof construction	116

	Trussed rafter detail	120
	Framing and covering details	120
	Laminated truss detail	124
	Flashings	124
7	**DOORS**	127
	Timber	127
	Doors	128
	Fire-resistant doors	131
	Sound insulation doors	132
	Door frames and linings	132
	Door furniture	136
8	**WINDOWS**	138
	Principles of framing	138
	Window analysis	139
	Opening lights	140
	Timber windows	141
	Metal windows	141
	Window surrounds	143
	Sliding sash window	144
	Glazing	145
	Double glazing	146
	Wall glazing	147
	Curtain walling	148
	System building	150
9	**STAIRCASES**	152
	Staircase layouts	152
	Design points	153
	Wood staircases	154
	Stone staircases	158
	Open-tread staircases	160
10	**PARTITIONS**	162
	Partitions	162
	'Eggbox' partition	165
	Dry lining	166
	Internal finishings	166

	Plastering	168
	Pre-mixed plaster	169
	Ceilings	170
11	APPLIED FINISHES	172
	Paints	172
	Reasons for painting	172
	Types of paint	173
	Preservative coverings	177
	Preparation of surfaces	177
	The paint system	178
	Paint application	179
	Painting defects	181
	Wallpaper	181
	Plastic wall coverings	182
	Standard colours	182
12	BUILDING MAINTENANCE	185
	New work	185
	Routine maintenance	187
	Older property: maintenance reports	187
13	BUILDING ALTERATIONS AND REPAIRS	193
	Dampness in buildings	193
	Dry rot	194
	Wet rot	195
	Woodworm	196
	Protected timber	196
	Shoring	197
	Underpinning	201
	'Saw-cut' D.P.C.	203
	Electro-osmotic D.P.C.	204
	Drying out	204
	Replastering	204
	Building measurement	205
	APPENDIX	212

FURTHER SOURCES OF INFORMATION 212
British Standards 212
British Standard Codes of Practice 213
Building Research Establishment Digests 213
Agrément Certificates 213
Government publications 213
Trade literature 213
RIBA Product Data 214
N.B.A. and 'Building' 214
Classification of information 214

INDEX 220

Preface

Building is a team challenge, and to meet this challenge, the clear communication of information is essential. This communication in the form of well-presented drawings and expert instructions, backed by a thorough knowledge of basic principles, is a vital link in the better understanding of the complete process of building. The whole field of building technology has become very complex and relies on an increasing necessity for specialization. At the same time the industry has passed through a period of great change by reason of the introduction of new techniques and the necessity to learn new ways of applying traditional knowledge. In building, the skills involved are those requiring an expert understanding of organization and management, a knowledge of the development of production techniques used in the manufacture of building components, and of course the experience and skill of the traditional trades. There are also the many techniques concerned with the process of site assembly and fixing of factory-made units. Invariably in such a field there will be many whose responsibility lies in one aspect of construction but who must understand the implications of their work against a background of general knowledge of the total problems involved.

In addition to the professional staff and the trained craftsmen with their specialized skills there is the need in the construction industry for technical staff both on the site and in the office. The technician's work involves production planning, method study, quality control, surveying, measuring, scheduling, estimating, supervision and setting out.

All who study building must have a sound knowledge of the principles of construction, and should be aware of the implication of new techniques. They must also have an appreciation of the skills called for in all aspects of the building process.

This book is an attempt to put very simply the general picture of present-day techniques through the principles of elementary building construction and services, covering as wide a field as is possible and practicable. At the same time the student is introduced to an understanding of the problems associated with the more complex forms of building technology.

It is hoped that the book will be equally useful to the trainee technicians and trade apprentices, and to students of architecture, surveying and structural engineering in the early years of their training. To anyone with specialist knowledge their particular field may here seem to be too elementary, and perhaps even superficial, but limits have to be set; and if the basic principles are clearly understood, the general properties of the materials grasped, then new developments and improvements will be more easily appreciated. Thus specialists in a particular field who desire a wider knowledge of the basic principles of other aspects of building work should also find the book useful and will be better able to pursue their particular studies against this background.

Hexham   HAROLD KING
January 1976

Author's Note to First Edition

This book was originally published as *Building Construction Illustrated* by Denzil Nield, A.R.I.B.A. It has now been thoroughly revised and expanded into two volumes in the light of current practice and development, by Harold King, M.A., A.R.I.B.A. Volume One deals with Structure in Building; including site investigation, foundations, walling, floors and roofs, windows, doors and stairs. Volume Two includes Services; covering cold and hot water supply; Heating and Ventilation; Electrical work; and external works and drainage; also Finishes, painting and wall-papering; and Building Maintenance, including defects, repairs and building measurement.

With regard to the material contained in this volume, the appropriate CI/SfB classification and a brief list of the relevant British Standards and Codes of Practice is given in the Appendix.

The Authors wish gratefully to acknowledge the help received from the following:

Peter Burberry, A.R.I.B.A., for those drawings retained from the original book, and to Christine Moore, Jean Turner and Hugh Williams for invaluable drawing office assistance in the preparation of the new diagrams. Also to Mrs E. M. Thomas and Karen Symonds, for careful work in translating manuscript to typescript; the British Woodwork Manufacturers' Association, for permission to use material from standard details, registered under the E.J.M.A. certification trademark; the Timber Research and Development Association; and to all the manufacturers who have supplied technical information.

HAROLD KING
DENZIL NIELD

May 1966

Author's Note to Revised (S.I.) Edition

This edition uses terms and symbols based on the SI system (Système International d'Unités). A table of the symbols used in this book is given on page xvii.

The text covers the Site; Load-bearing Construction; Framed Construction; Floors; Roofs; Doors; Windows; Staircases; and Partitions, and now includes chapters on External Works; Finishes; and Building Maintenance and Repairs, which were originally in Volume Two.

The subject matter has been brought up to date, in line with current practice and recent changes in legislation affecting building.

I wish gratefully to acknowledge the help given by D. M. Morris, Dip.Arch., A.R.I.B.A., for preliminary work on the manuscript for Paths, Pavings and Fencing, and A. Ashton, for practical comment on Building Maintenance regarding the material originally in Volume Two. Also to K. R. Mitchell, for advice on the chapters dealing with the Site and Survey, and to Adrian Napper, B.Sc., C.Eng., M.I.C.E., for advice and comment on the Structural Engineering content of this edition. Also to J. M. Harrison, for helpful comment on the work in Slating and Tiling.

My thanks also to Mr John Merriman for help in bringing the chapter on 'Applied Finishes' up to date.

The building regulations referred to are *Building Regulations, 1972*, together with amendments up to the time of printing (First Amendment Regulations, 1973; Second Amendment Regulations, 1974; and Third Amendment Regulations, 1975). These regulations are current in England and Wales (except London). For Scotland, refer to *Building Standards (Scotland) Regulations*, and for London, *London Building (Constructional) Bylaws*. The regulations for London and Scotland have similar objectives to the England and Wales regulations but differ slightly in coverage and detail.

Note that Part III of the Health and Safety at Work etc. Act, 1974, contains certain powers with regard to building regulations in general and this will have a very important and increasing influence on future building legislation, and thus eventually building techniques.

January 1976 HAROLD KING

The Metric System

SI units and symbols:

Length	metre	m
	millimetre	mm
Area	square metre	m^2
Volume	cubic metre	m^3
Frequency	hertz	Hz
Mass	kilogram	kg
Density	kilogram per cubic metre	kg/m^3
Force	Newton	N

A Newton is a unit of force which applied to a mass of one kilogram gives it an acceleration of one metre per second per second, i.e. $N = kg\ m/s^2$

Pressure, stress	Newton per square metre	N/m^2
Temperature interval	degree Celsius	degC: $°C$
Coefficient of heat transfer	watt per square metre degree	W/m^2 degC

CHAPTER ONE
The Site

Choice of site
The assessment of the suitability of a plot of land for building development is a matter which requires considerable skill and experience. The person who intends to develop the site must be satisfied that the topography of the land is satisfactory; that the sub-soil is adequate to support the intended structure; that there is good access from suitable roads; that all necessary services such as water, drainage, electricity, gas and telephone are available; that the aspect in relation to the sun is correct for the intended use of the building; and for certain sites, that the views both outward and inward are good. In addition there should be no disabling easement such as Rights of Way across the land by others. It will be readily appreciated that the ideal site is hard to find and so a skilful compromise of conditions will probably have to be made.

Before building starts, Local Authority approval will have to be obtained under the Building Regulations and the Town Planning Acts, and the first step is to make a survey of the site. Certain buildings will also have to satisfy the requirements of other legislation relative to buildings. For instance, the Factories Act, the Fire Precautions Act and the Health and Safety at Work Act.

Site survey
Diagram 1 shows a hypothetical site and gives the information which should be included in a field survey. The site will first be surveyed and levelled; topographical details being measured by means of metal chains laid down on the site and used as base lines for measurement taken at right angles to them and known as 'off sets'. Diagram 2 shows a typical specimen page from a Field Survey Book to illustrate the technique. The term 'levelling' refers to the use of an optical instrument incorporating a system of spirit levels and a telescope with a sighting device so that comparative readings may be taken on a horizontal plane. These readings are known as 'spot levels' and enable the contours to be

2 · BUILDING TECHNIQUES

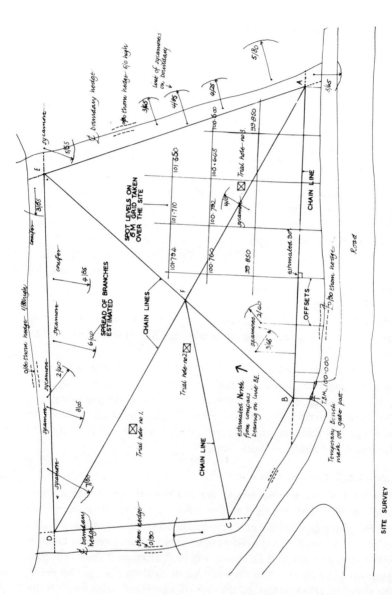

Diagram 1　Information obtained from a site survey

THE SITE · 3

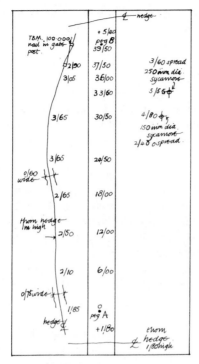

PAGE FROM FIELD SURVEY BOOK TO
SHOW CHAIN LINE A B

Diagram 2   Field book

plotted on the site plan. Most levelling instruments have automatic or semi-automatic devices which enable them to be set up quickly and accurately on the site and are thus known as 'autoset' or 'quickset' levels. The chain survey procedure described is especially suitable for 'green field' surveys, but where difficult or large areas need to be surveyed, techniques using optical instruments such as the theodolite are to be preferred. This kind of survey will probably be undertaken by specialist surveyors.

Setting out

The setting out of the foundations and walls is done by driving wooden pegs into the ground on the intended line of the foundation trenches and then fixing profile boards as shown in diagram 3. Cords are stretched from board to board to come over the pegs fixing the position of each wall. Marks are drawn or cut into the top of the profile boards

4 · BUILDING TECHNIQUES

so that the cords can be removed when the bricklaying starts but can easily be put back for setting out purposes.

Profile boards will be erected at all angles and cross walls and later the lines for the outer edges of the foundation will also be marked on the boards. A right angle can be set out by using a builder's square. This consists of three lengths of timber fixed together to form a triangle with sides of 3 units, 4 units and 5 units long. After one corner has been set out in this way it should be checked by measuring within the angle a triangle with sides which are multiples of 3, 4 and 5.

The 'datum line' in the building should be given by the Architect in relation to the level of a given point defined on the site plan. This is then set out on the site with a levelling instrument. The 'datum' is a level marked on the sections of the drawings from which all heights and depths are marked in figures. The 'datum' is usually taken as the surface of the finished ground floor level, abbreviated on working drawings to F.F.L. A peg is driven into the ground so that the top is at the 'datum level' and it is then good practice to fix the tops of all profile boards at a convenient measurement above this so that the levels of the bottoms of trenches and the top of the foundation concrete can be set out by 'siting through', using boning rods of a given length in the manner described in the chapter on Drainage in Volume II.

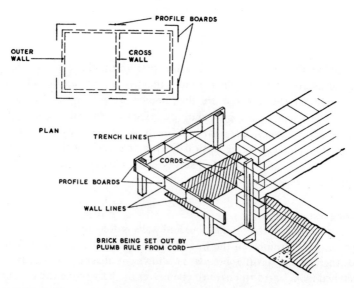

Diagram 3   Setting out

THE SITE · 5

Removal of top soil
Before the foundation trenches are excavated, the surface vegetation, roots, plants and shrubs, and usually all the top soil, probably to a depth of between 150 and 300 mm, will have to be removed from the area of the site to be covered by the building. The effective removal of the turf and vegetable matter is required by the Building Regulations to ensure that the ground upon which the structure will be erected will be sterile. The top soil is valuable for subsequent use in the garden layout, and is best removed to a 'spoil' heap conveniently placed.

Site drainage
The Building Regulations also require that the sub-soil of any site that is to be used for building must be effectively drained; thus if the natural drainage of surface water through the ground is not sufficient, a line of agricultural drain pipes can be laid on the uphill side of the site to intercept ground water that would otherwise flow towards the building. These pipes may be porous earthenware (B.S. 1196), or porous concrete (B.S. 1194), and should be specified to be 'non-choke' with a flat solid base. The upper part only is porous and the pipes fit tightly together to prevent debris or roots entering the pipe lines. The pipes are laid in gravel or hardcore packed trenches. The land drain may be discharged into a nearby ditch or stream, but if this cannot be done a catchpit should allow the silt to settle and a trap must be constructed between the land drain and its connection to the main drainage system.

Excavation
When the setting out is completed and the profiles are in position, excavation of the trenches for the foundations is started. The width of the trench is read off from the profiles and the depth of the trench depends on finding a suitable sub-soil to give a firm bearing. The foundation concrete must also be below the depth at which it will be affected by seasonal movement of the sub-soil, and in Britain a depth of 750 to 900 mm would be satisfactory. It is now common practice to use earth-moving machines for the excavations on all contracts except perhaps on an isolated small site. Machines which will be found in most common use will be a hydraulic digger for trenches, a tractor shovel for reducing levels by excavation and a dumper for transporting spoil on the site. Often the digger and shovel will be mounted on the same vehicle which may be fitted with either wheels or tracks. When the spoil has been tipped by the digger it can be lifted by the tractor shovel and tipped into the dumper to be carried away and discharged wherever it is

6 · BUILDING TECHNIQUES

required for making up levels on the site. Spoil which is re-used in this way must be placed in layers, well rammed and allowed to settle for many months so that it will not subside later if built upon. Typical machines are shown in outline in diagram 4.

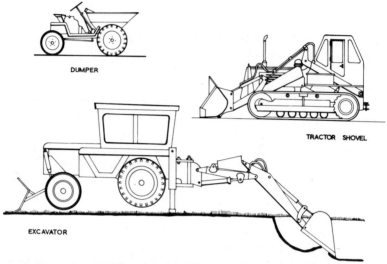

Diagram 4   Excavation equipment

Safety in excavations

Accidents in excavations are frequent and include a high proportion of fatalities. These accidents are just as likely to happen in relatively small excavations. A cubic metre of soil weighs more than a tonne and falling through only a short distance, even half a cubic metre of soil is sufficient to crush and kill a workman. Thus great care should be taken to support the excavations adequately.

The particular circumstances of each site have to be considered before deciding how the sides of the foundation trenches can best be supported and so it is difficult to legislate for safety. In soft clay sub-soils the sides of the trench will only 'stand up' for a shallow depth and so close timbering or metal sheeting will be required as shown in diagram 5. For very soft clay the sheeting must be driven ahead of the excavation since it would probably not be possible to excavate to the full depth of the foundation. The trench sides are more likely to collapse if spoil is dumped near to the edge of the excavation. Thus, all spoil heaps should be kept at least the depth of the trench away from

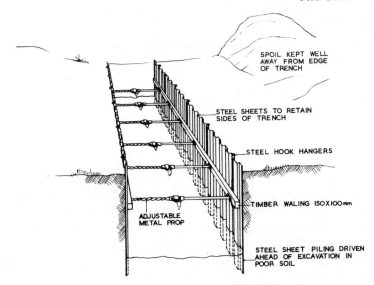

Diagram 5   Support to excavation trenches

the edges. Stiff or gravelly clay will stand up unsupported for a considerable height and so it is a great temptation to take risks with this type of soil, but the stability is misleading since what appears to be a good face is, after a time, liable to collapse without warning. It will be appreciated that the question of supporting the trench walls is a matter which requires considerable experience in order to determine the best method to use. On large contracts the support will be detailed in the design office of the consultant engineers.

Where timber is used, it should be tight against the face of the soil and the wedges should be redriven regularly, watch being kept to see that there is no erosion of the ground behind the sheeting since this will cause slackening and instability of the timbering. It is said that it is a favourite pastime in Britain to watch other people working — so barriers should be provided to prevent injury to the curious. Excavated material should not be balanced on struts and great care must be taken when dismantling the supports.

CHAPTER TWO
External Works

This section of the work deals with the construction and detailing of the external works associated with a building. It is concerned primarily with the area immediately surrounding the building and this includes the work in connection with minor roads such as may be found on a housing site, the provision of access roads, the surfacing of parking areas, the provision of paths and paved areas and the construction of fences and gates.

Estate roads
Housing estate roads will have to be constructed to the approval of a Road Surveyor to the standards set down by the Local Authority, since the future maintenance of the road once it has been 'adopted' becomes the responsibility of the Local Authority. Thus the Road Surveyor of the Local Authority should always be consulted at the design stage, and the following definitions will be helpful in understanding his terminology:

Road: a way for vehicles; note that the term 'roadway' should not be used.

Service road: a subsidiary road connecting with a principal road and providing service access to adjacent buildings.

Footway: the portion of the road reserved exclusively for pedestrians. Note that the term 'pavement' in this context is incorrect, and that a 'footpath' is not the same thing, being a way across open spaces.

Carriageway: the portion of the road reserved exclusively for vehicles.

Pavement: This has two meanings:

(*a*) a general term for the paved surface of either a carriageway or a footway, and

(*b*) the term applied to the whole construction of the road above the foundation.

A road may either have a flexible, granular (hardcore) base or a rigid monolithic (concrete) base. Most estate roads will be of the flexible type of construction since concrete is now reserved for sites where difficult sub-soil conditions exist. The final strength of a road depends upon the strength and condition of the sub-soil (sub-grade) beneath it and much experience is required to determine what preparation, for example, in the form of extra drainage, the sub-grade of a proposed roadway will require before the road is laid. The strength of the sub-grade can be determined scientifically by the use of the California Bearing Ratio testing technique. Tests on the sub-grade are made either *in situ* or in a laboratory. The bearing strength of the sub-grade is worked out from the results and then the thickness and quality of the base materials to be used to make up the road can be determined by reference to a graph. The graph also relates to the use the road will have, based on the average number of commercial vehicles that are expected to use the road per day. The expense of testing, however, would be prohibitive for small schemes and here the advice of the Local Authority Surveyor should be taken. The following is a suitable specification for an average estate road:

The widths of carriageways and footpaths should be in accordance with the latest Ministry recommendations and visibility splays must be provided at all road junctions to the requirements laid down by the Local Authority. The longitudinal gradient of carriageways and footways should not be less than 1 in 150, and preferably 1 in 120, to ensure satisfactory drainage. Carriageways should have an average camber from channel to crown of 1 in 30 and footways a cross fall from back of path to kerb of 1 in 24.

Carriageway foundations must be approved hardcore, laid, rolled and blinded, to a compacted thickness of 225 mm on a sub-base of well-rolled fine graded quarry waste not less than 75 mm thick. In the case of minor cul-de-sac roads or service roads the depth of hardcore or stone may be reduced to 150 mm. The carriageway could be surfaced with 50 mm finished thickness of 38 mm nominal size bitumen macadam to B.S. 1621 with a wearing coat of Fine Cold Asphalt to B.S. 1690 rolled to 20 mm finished thickness.

Footways could consist of a base of approved hardcore of a finished thickness of not less than 75 mm surfaced with 25 mm thickness of 15 mm bitumen macadam with a topping of Fine Cold Asphalt of 12 mm finished thickness.

Kerbs should be hydraulically pressed concrete kerbs 250 mm deep × 125 mm wide to the current British Standard. They should be laid on a bed of concrete not less than 325 mm wide × 100 mm thick, the front of the bed being flush with the face of the kerbs, and haunched up at the back with concrete to within 75 mm from the top of the kerb. The back of the footway should be supported by 150 mm × 50 mm concrete edging bedded on and haunched with concrete. Diagram 6 illustrates a typical cross section of an estate road.

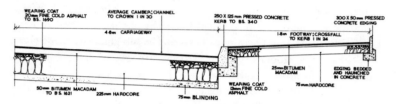

Diagram 6   Typical section of an estate road

Road drainage

Adequate provision must be made by gullies and drains for the disposal of surface water. A general rule is that a road gulley should be provided in each channel at maximum 6 m apart, the setting out being done from each low point determined by the contours of the ground. For Housing Estate layouts the Civil Engineering Contractors who will construct the road, and who will probably, at the same time, lay both the surface water and foul water drains under the road should be provided with a schedule and sections showing the depths and gradients of the drains. Gulley pots may be of stoneware or concrete and should be 450 mm in diameter by 900 mm deep, and gulley gratings should be of cast iron to an approved pattern. Two or three courses of engineering bricks set in cement mortar are normally laid on top of the gulley so that the cast-iron gratings and frame can be set at the correct level in the road channel. There are several patterns of road gulley gratings. The grating is usually hinged on to the frame and its weight is calculated according to the wheel load it is expected to withstand bearing in mind the type of road. Diagram 7 shows a typical road gulley and cast-iron frame and grating suitable for estate and service roads. An alternative type of gulley cover is also shown with a side inlet suitable for setting in the kerb.

Manholes may be of brick, or of precast concrete sections, with heavy-duty cast-iron covers to withstand the wheel loads from vehicular traffic. A 1 m diameter manhole would be suitable up to about 2 m deep, and below this, a taper shaft would be fitted. The manhole should

EXTERNAL WORKS · 11

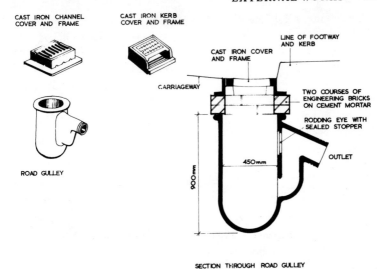

Diagram 7   Road gulley

be surrounded by 150 mm concrete. Building drainage should be carried out in accordance with the recommendations of British Standards Code of Practice 301; B.S.C.P. 308 covers the drainage of roofs and paved areas.

Where the rainwater disposal is completely separate, the connection between the rainwater pipe and the branch surface-water drain may be made by means of an untrapped rainwater shoe, although a trapped gulley is preferable where there is any danger of foul air entering the system from an external source. A typical rainwater shoe is shown in diagram 8. Where it is not practicable to connect a branch surface-water drain to the main system of surface-water disposal a Local Authority may allow the branch drain to be connected to a soakaway, provided the ground is suitable. A typical soakaway is shown in diagram 8.

Edging
The edging material used at the immediate perimeter of a building, that is, the detail of the junction of the building with the ground, is of importance both from the visual and from the constructional aspect. Many designers like to see surrounding grassed areas taken right up to the building, but this causes difficulties in keeping the grass neatly cut and so a mowing edge of say 600 mm x 300 mm x 38 mm thick concrete paving slabs is a good detail at the base of the wall to allow the grass to

12 · BUILDING TECHNIQUES

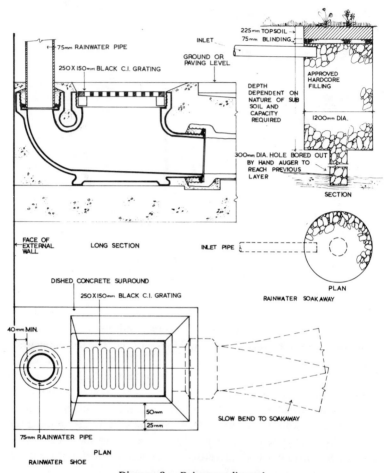

Diagram 8   Rainwater disposal

be cut to a firm edge. For this reason also, all turfed or grassed areas should be finished at a level about 50 mm above surrounding hard surfaces. As an alternative to the slabs, a small trench 150 mm deep and 225 mm wide filled with gravel will be satisfactory and will also help to drain the lawn. Both these details are illustrated in diagram 9.

Where flower beds are arranged adjacent to the building, care should be taken to keep the soil at least 150 mm, and preferably more, below the damp-course level and to make sure that plants and particularly creepers do not cause damage or bring dampness into the building. A brick edging set on a small concrete bed will prevent the spill of soil from a flower bed on to an adjoining path, as shown in diagram 9.

EXTERNAL WORKS · 13

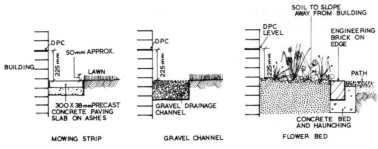

Diagram 9   Edging details

Paved surfaces
The choice of materials suitable for use in paved areas requires careful consideration. It is important to understand the function of the paving in order to enable a correct choice to be made. The main purpose of paving is to form a solid carriageway for pedestrians or vehicles. This requires careful choice and laying of base and surface to withstand the loading involved. The surface should be well drained and should thus have carefully planned falls to carry the surface water away. Paving slabs may also be used as floorscape to give pattern and a sense of direction to the layout. Pedestrians will tend to avoid uneven cobbles and may prefer to keep to the smoother pavings. The following is a list of the most commonly found materials used for paved areas:

(1) *Tarmacadam:* a wide range of tar-based materials are used for surfacing both vehicular and pedestrian paved areas. They consist of an aggregate mixed before delivery under controlled conditions with tar. The material is laid and rolled on a prepared surface of well consolidated and rolled fine hardcore in one or two coats. Bitumen macadam has similar properties but uses bitumen and not tar as binder.

(2) *Coloured macadam:* wearing courses using coloured aggregate or a coloured binder are available for footways or playgrounds, dark green or red being the most usual.

(3) *Asphalt:* asphalt paving has a much finer surface texture than tarmacadam and as it sheds rainwater more easily it can be used in areas which are almost level. Where falls are not a problem, tarmacadam would have the advantage of a greater surface grip. The main forms of asphalt in common use are hot rolled asphalt and fine cold asphalt. Hot and cold asphalt are laid in one course only on a variety of bases but usually as a wearing coat over bitumen macadam or tarmacadam. The cold rolled material is by far the most popular.

(4) In situ *concrete:* this is an inexpensive method of paving but of little aesthetic value. It should be laid in bays to allow for expansion and contraction. The thickness of the slab will be 50, 75 or 100 mm and the maximum area that can be covered will be dependent on the quality of the concrete mix. The joints can be so designed as to be useful drainage channels. Cobbles or granite setts are often used to divide the concrete into bays and allow for expansion at the same time.

(5) *Gravel or shale:* this is obtained from pits or river beds and varies in colour according to the locality of its source. It is laid loose and should be well rolled to form a surface suitable to walk upon. The layers can vary from coarse hardcore to a fine topping, depending on requirements. It is an inexpensive method of providing a dry pathway and has traditionally been used as garden paths retained between boarding edges. It requires careful maintenance by a keen gardener, however, to keep down the weeds.

(6) *Pre-cast concrete:* concrete slabs are widely used and can be obtained in a large range of sizes, shapes, textures and colours. Slabs 300 mm x 300 mm, 300 mm x 600 mm, 600 mm x 600 mm and 600 mm x 900 mm and in thicknesses 38 mm or 50 mm, are typical of the range. The most common forms of slab are rectangular and square but hexagonal slabs are also used. There are three methods of laying concrete pavings, according to the type of wear expected. Where wear is known to be heavy and freedom from movement is essential it is necessary to lay the paving on a solid bed at least 25 mm thick of semi-dry 1 : 4 cement/sand mortar.

The ground should first be made as smooth, compact and level as possible. The joints should be set 6 mm or 12 mm wide and these can then either be left open or pointed up. For normal foot traffic, the slightly cheaper method of setting the slabs in mortar dabs can be used. The ground should first be levelled as before and then a fine bed of sand 50 mm or 75 mm thick laid and rolled. The number of dabs depends on the size of the slab, one being sufficient for a 300 x 300 mm paving, and five for a 900 x 600 mm slab. Where only casual foot traffic is anticipated, concrete slabs may be laid direct on a 25 mm bed of fine sand. Each slab is 'settled' in the sand and laid with open joints. With this method some movement or settlement of the slabs must be expected.

(7) *Brick paving:* engineering bricks manufactured as pavers form an extremely interesting and durable surface. They are manufactured by wire cutting and pressing, but being impermeable and vitrified they tend to have a slippery surface. It is therefore preferable to use the diamond chequer or panelled variety.

Plain bricks have also been used extensively for paving in the past, in particular in Scandinavian countries. They are also popular in Britain but have several practical drawbacks. A moss or lichen may grow over them making the surface dangerously slippery, and in fact some types of hard, smooth bricks become very slippery even after only a short rain storm. Brick pavings suitable for pedestrian traffic can be laid quite satisfactorily on a sand or 1 : 4 cement/sand bed. For vehicular traffic, a concrete foundation should be used.

(8) *Pre-cast exposed aggregate concrete slabs:* this form of paving is made in similar manner to the slabs which are now used for the cladding of buildings. Not all aggregates are suitable, sharp stones being obviously unsuitable for areas to be walked upon. But the multitude of textures available in this kind of paving make it a very attractive material for formal layouts. Aggregates range from a small chipping to the size of large cobbles. The slabs are laid in a similar manner to the concrete pavings previously mentioned.

(9) *Cobbles:* these are mainly used for their decorative effect or as a way of keeping the pedestrian to fixed routes. The best cobbles are procured from river beds and beaches. They vary from spherical to kidney-shaped samples and in sizes graded from about 40 to 100 mm in 'diameter'. The laying can be done in a random form; roughly coursed, or deliberately patterned. The cobbles can be set in 1 : 5 cement mortar to half their diameter, laid dry and afterwards watered through a watering-can using a rose to give a fine spray. It is usual to find that the finished level of a cobble bed is slightly above neighbouring surfaces.

(10) *Granite setts:* these are a very durable form of paving, used now in small areas and extensively for trim to other forms of paving or grass areas. Particularly used in the form of drainage channels. Many of the setts used in modern layouts are obtained from old city streets, now being renewed with asphaltic surfaces. The setts are best laid in a dry cement

mortar on a concrete bed and afterwards watered as described for cobbles.

Diagram 10 shows the methods of laying tarmacadam; gravel; pre-cast concrete slabs; cobbles; and granite setts on suitable foundations.

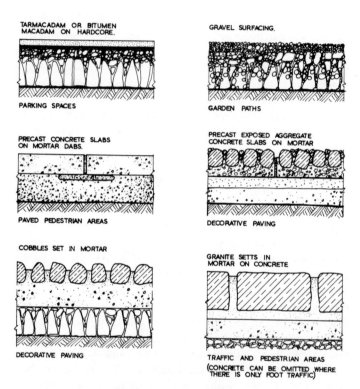

Diagram 10   Types of paving

Kerbs and trim

Kerbs are generally pre-cast in concrete, common sizes being 300 × 150 mm and 250 × 125 mm. The upstand kerb is the usual way of finishing the edge of an urban or estate road. Diagram 6 shows the form this normally takes with the channel falling to the road gullies. A thin pre-cast concrete round edge strip 125 or 150 mm deep and 50 mm wide is a useful edging for pavements and paths. The concrete trim should be set in a small concrete foundation with the rounded edge just projecting above the height of the adjoining surfaces.

Fences

The most widely used materials are timber, concrete, wrought iron, stranded wire and chain link, these can be compared in diagram 11. Stranded wire used either with timber or concrete posts is a very inexpensive form of fencing and is therefore widely used on housing estates, but unfortunately has little aesthetic value. In this type of fence where the strands are in tension, straining posts must be introduced at ends and corners and any long straight sections of fencing should have straining posts at least every 70 m.

Other forms of wire fencing which give varying amounts of enclosure are woven wire and the popular chain link. These also require straining posts. An example of a 2 m high fence is shown in diagram 11.

The concrete fence is generally heavy in appearance and lacks the weathering qualities of its natural timber counterpart. It is, however, extremely durable and requires no maintenance. The construction usually takes the form of concrete posts at approximately 1·8 m centres linked by pre-cast planks which are located between the posts, producing a ship-lap boarding effect.

Wrought iron is the traditional material for fencing (or more properly railing) and very many fine examples of the blacksmith's art can still be seen surrounding seventeenth-century and eighteenth-century buildings. Unhappily the modern counterpart when used decoratively for screens or gates does not always maintain this fine tradition.

Timber offers the greatest variety of forms and effects to the designer, and being a natural material fits well with any landscaped layout. There are many types of fence available as stock items in any joinery manufacturer's yard. Timber fences range generally in height between 1 and 2 m, centres of posts between 1·8 to 2·4 m depending on type, and sections of posts between 75 x 75 mm and 150 x 150 mm.

Posts can be fixed in several ways. Setting in concrete provides a very firm fixing, but tends to rot the timber at point of entry. Alternatively the post can be driven into position. This method tends to damage the head and encourage the entry of water.

The third method is to wedge the post with large stones in a hole, placing more small stones around the base before back-filling and well ramming the surrounding area. This ensures good drainage around the base of the post. To increase the life of a post it is preferable to use hardwoods, wherever possible. Whichever method of fixing is used, a fence or gate post should be in proportion to its length of $\frac{1}{4}$ below ground and $\frac{3}{4}$ above ground, with a minimum of 450 mm below ground.

18 · BUILDING TECHNIQUES

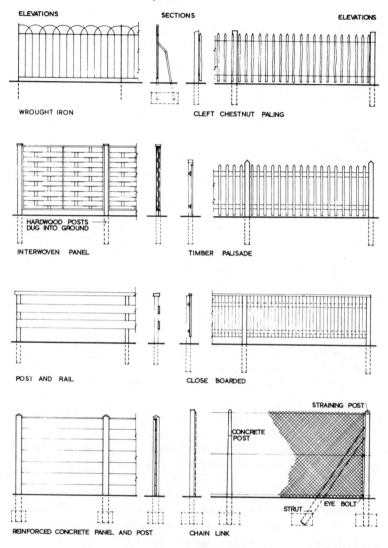

Diagram 11 Fencing

The most durable species of home-grown timbers for use in fencing are Oak, Larch and Chestnut; any of these can be successfully used without any form of preservative. Many other home-grown timbers can be quite adequate if treated at regular intervals. The timber should be pressure impregnated by preservative before the erection of the fence,

EXTERNAL WORKS · 19

and then brush or spray treatment as required at annual intervals. It should be noted that the ends of the posts to be inserted into the ground should receive special attention. In the case of the 'no maintenance' species mentioned above, preservative will not penetrate very well. It is therefore desirable to coat the post below ground with a tar material, thus forming a waterproof coating.

Capping of fencing posts and vertical members of fencing is desirable to protect the end grain and thus prolong the life of the fence. Where this is not possible, a weathering face should be formed to take water quickly away. Note that galvanized nails must always be used to avoid black stains. The various types of timber fencing are listed below and a selection of the patterns is illustrated in general outline in diagram 11.

(1) *Cleft chestnut paling:* this is used for temporary enclosure, but nevertheless has a certain charm in its appearance and is often used on a more permanent basis in country areas. The chestnut pales are strung along twisted wire between the supporting posts, as shown in diagram 11.

(2) *Interwoven:* this type gives good privacy and protection from prevailing winds. It is very well suited to a suburban garden (many types being advertised as 'peep-proof'). There are two main types of panel — the Overlap and the Interlace.

(3) *Close boarded:* a fence which is strong and durable, gives good privacy and protection from wind. The most popular type is formed on two triangular-sectioned horizontal bearers spanning from post to post on which the pales are nailed with about 20 mm overlap.

(4) *Palisade:* a neat fencing allowing a view through at the same time as acting as an efficient barrier. It is particularly attractive when painted white, as is the usual tradition.

(5) *Wattle hurdling:* this is a traditional form of fencing which is now becoming popular again. It gives complete privacy and acts as a good windbreak. Separate hurdles are made in about 2 m lengths from Hazel wattle which weathers pleasantly brown. For long life, it should be held just clear of the ground.

(6) *Post and rail:* an infinite number of combinations can be achieved with the post and rail principle. The popular modern idiom is to use deep rails, which give a very strong horizontal effect when painted white. Sizes of timbers vary according to the desired effect.

(7) *Louvred fencing:* this type of fence is a variation on the post

and rail construction and has become popular where fencing is required to restrict, but not to completely cut off, the view through.

(8) *Wrought-iron fencing:* diagram 11 shows a typical pattern and it will be seen that the wrought-iron railing in this form is both protective and decorative. The wrought iron is produced to laboratory specification and is protected by primer before delivery.

Gates

These are manufactured in a great variety of sizes and forms. The main factors to be considered when choosing or designing a gate are:

(1) Make sure that the material is not incongruous with the material of the adjacent fence or wall.
(2) Keep the weight of the gate as small as possible without impairing the strength.
(3) Choose the furniture carefully. The traditional five-barred gate, shown in diagram 12, is hung on iron hooks with large strap hinges and is a form which has stood the test of time.

The notes regarding the fixing of fence posts are also relevant to gate posts, bearing in mind additionally that the hanging post must be strong enough to withstand the cantilever action of the gate and the clashing post securely fixed to prevent it from being shaken loose. Gate posts range in section from 75 × 75 mm for a small garden gate to 225 × 225 mm for a large field gate.

Diagram 12 also illustrates a pair of 1·8 m high framed, ledged and braced doors, suitable for use in a boundary wall.

Boundary walls

The range of materials available for external walls is considerable and offers a great variety of colours and textures. The factor to be taken into consideration is that both sides of the wall are exposed to the weather, and so the use of the correct mortar is very important, as is the design on the coping or the top surface of the wall.

The foundations should be very carefully considered, and also the thickness in relation to the height, particularly if the wall is used to retain earth on one side, as is often the case when walls are used at changes of level. Traditional wide-strip foundations are suitable for normal conditions of loading.

Brick boundary walls

Bricks that weather well and have good frost resistance should be chosen for boundary walls.

EXTERNAL WORKS · 21

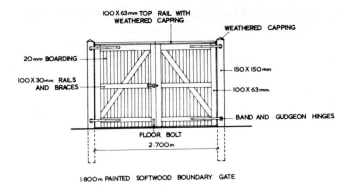

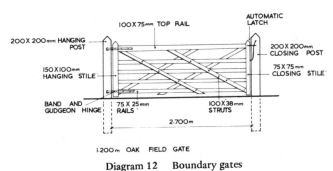

Diagram 12   Boundary gates

Although the wall is exposed on both sides, and there are no finishes to be affected by rising damp, it is nevertheless advisable to insert a horizontal damp proof course, to keep the wall as dry as possible and prevent harmful sulphate attack from the damp sub-soil.

Half-brick thick walls are not very good but they can be used up to a height of say 1·5 m, provided that piers are constructed at no more than 2·5 m centres. One brick thick boundary walls can be built to the same height without piers. Notes on bonding of brickwork, mortars and jointing and pointing techniques in respect of general construction are given on pages 45–50 and are equally applicable to boundary walling.

Stone boundary walls
The choice of type of stone and techniques of building are many and varied and there are many excellent examples of varying traditional methods.

Dry walling is still used in some parts of the country but requires great skill in choosing and placing by the mason. The wall is usually formed with uncoursed random rubble on average 450 mm thick with a throughstone placed at least every square metre.

Other types of commonly used stone walling are as follows:

Coursed random rubble — the stones are used as found, but during the course of construction the units are brought to a level bed perhaps four times in a 2 m wall. (See diagram 13).

Regular coursed rubble — the stones are carefully chosen to form approximate courses.

Square ashlar — this is a carefully prepared stone, generally using softer limestone or sandstones, dressed square and laid in courses. Thickness of bed can be reduced considerably over that of the rubble walls, and it is not uncommon to find walls of say 1·5 mm having a bed of only 150 mm.

Concrete block boundary walls

The use of concrete both *in situ* and in block form as a walling material has increased enormously over the years, and consequently great advances have been made in the development of surface treatments. The appearance of natural concrete from a smooth shutter is very harsh, dull in colouring, prone to excessive staining and bad weathering, consequently it can easily spoil the appearance of a building when thoughtlessly used in boundary walling. Care must therefore be taken in the choice of aggregate which controls the colour and in the surface treatment which dictates the texture. The cheapest form of concrete walling is the mass-produced block, generally 450 x 225 mm in size obtainable in several thicknesses, 150 or 225 mm being common. Blocks can be obtained with an exposed aggregate finish, the texture being varied by the size of the aggregate used. Decorative pierced blocks can also be obtained where special effects are desired. Several firms now manufacture a range of this type of block to form boundaries where a partial view through is desirable.

Changes of level

Often the function of a boundary wall is to emphasize or protect changes of level. Diagram 13 illustrates this, and gives an example of a boundary wall in brick and in stone.

The brick example would be built in engineering bricks and shows bull-nose bricks used as copings. The steps and seat are pre-cast reinforced concrete slabs 65 mm thick and built 225 mm into the wall. The wall is built 'Flemish Garden wall bond' having one header to three stretchers in each course. Note the drainage trench filled with rubble to

EXTERNAL WORKS · 23

allow water from the sub-soil at the higher level to escape through weep holes in the small retaining wall. These weep holes are open perpendicular joints, every fourth stretcher one course above the lower terrace level. Note also the use of a brick-on-edge inverted 'arch' forming the curved 'battered' base to the larger retaining wall and linking the brick paving at the lowest level. The paving is of 50 mm pre-cast concrete slabs on a 25 mm bed of cement or lime mortar.

The stone boundary wall is built of coursed rubble. The retaining wall is shown battered to increase in thickness from 300 mm at the top to 450 mm at the base and is built of a plain concrete strip foundation although traditionally large stones, equal to twice the thickness of the wall, would be used as footing at the base. Note the back-fill of rubble to allow sub-soil drainage to the lower level. The water is allowed to

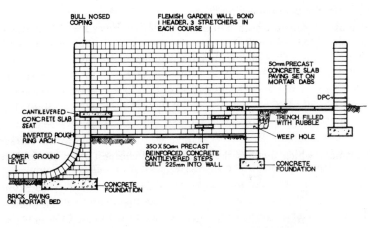

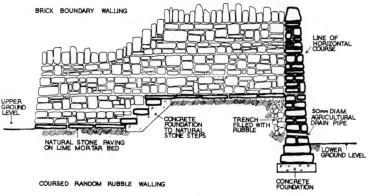

Diagram 13   Boundary walls

drain away through 75 mm diamter agricultural drain pipes set in the wall at 2 m centres.

The paving is of natural stone slabs laid on a bed of cement or lime mortar.

In the building of the walling, the larger stones are bedded flat and if necessary wedged up with small pieces of stone. No attempt is made to form vertical joints, but the work is levelled up every 300 to 450 mm to form a horizontal joint which goes right through the thickness of the wall. The wall is normally jointed in a weak lime cement sand mortar, but if the stones are very carefully interlocked and bonded by a skilful mason such boundary walls can be built 'dry', i.e. without mortar.

Copings

The weathering on the top surface of a free-standing wall should be very carefully considered. It is essential to avoid water penetration from the top which will quickly result in failure from frost attack. If walls built from a small unit such as brick do not have a coping, they require very regular maintenance in the form of re-pointing. However, this may be acceptable where a neat unobtrusive finish such as brick on edge is desirable and provided that the wall is built in a sheltered location. The provision of a coping is to be preferred and care must be taken in the choice of the correct material. Its purpose is primarily to prevent water penetration, but by projecting the coping, it also serves to throw water clear of the wall surface. Few materials used for copings are completely impervious, so it is also necessary to make provision for a damp-proof course just below the coping. The most common coping materials are: clayware, natural stone, slate, precast concrete and metal (non-ferrous).

Clayware: normal clayware copings are manufactured in two forms, weathered and saddleback. Faience may also be obtained in a large variety of forms and colours, although bright glazed colours often clash with surroundings and should be avoided. Clayware copings should be bedded in a 1 : 3 cement mortar.

Stone: the forms follow similar lines to the weathered and saddleback profiles of the clayware variety. Stone is more expensive but is often worthwhile where the design must match existing work. The stone chosen should be low permeability.

Slate: this is a dense form of stone completely impermeable to moisture and therefore an ideal coping material. It can be used in thin sections and consequently gives a very neat finish to the top of the wall. The base of a slate coping slab is scored and ready to be bedded in a cement mortar of 1 : 3 mix. Being thin in section, it is fairly easily displaced and therefore should be cramped and dowelled to the top of the wall. It is also desirable to join the slabs end to end with a halved joint and seal the joints with mastic. This is an expensive technique but makes a first-class coping.

Pre-cast concrete: the coping can be cast to any suitable profile and by the use of carefully chosen aggregates can be made to a variety of textures. The copings should be keyed by use of frogs and set in 1 : 3 cement sand mortar. Good curing of the slab is necessary as considerable shrinkage takes place on setting.

Metal copings: these are often used where a neat trim line is desirable to cap the wall. The metals in common use are aluminium, copper and zinc and they can be either fixed directly to the top of the wall or formed around a timber or concrete coping. A good drip should be formed and great care must be taken in the choice of metal for fixing screws, as electrolytic action between incompatible metals can seriously shorten the life of the coping. Diagram 14 illustrates in typical outline the copings mentioned above.

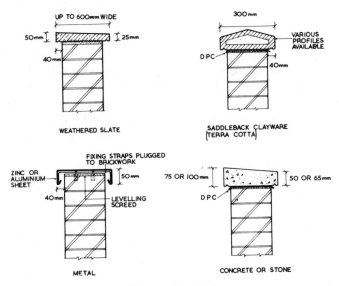

Diagram 14　Coping

CHAPTER THREE

Load-Bearing Construction

Walling
The designer must choose the method of wall construction most appropriate to the use of the building. The choice is between load-bearing units; monolithic; or framed construction.

Where load-bearing construction is used, the wall will be built of small units such as bricks, or concrete blocks, or alternatively if monolithic construction is favoured, the wall can be formed in solid concrete which will be poured between special shuttering and reinforced with steel. The third alternative is to construct a framework of structural steel, reinforced concrete, or timber, using columns and beams and leaving the space between the framing to be filled in with glass or solid panels to keep out the weather and provide insulation. Diagram 15 illustrates these alternative methods.

Load-bearing brick walls require to be buttressed at intervals or braced laterally by floors to prevent buckling. Walls at right angles, such as end or partition walls and solid concrete floors and flat roofs, do this very well. Wooden floors and roofs are less effective in this respect and care should be taken to fix the ends of the joists by means of wrought-iron straps or by cutting the brickwork and building up right against the timber. This latter technique is known as 'cutting and pinning'.

From a study of the various methods of construction it will be seen that bricks or concrete will absorb and hold water as a 'sponge' until the rain stops and the material dries out. For this reason, walls of soft brick or porous stone were traditionally very thick and even then were lined on the inside with lath and plaster, kept away from the wall by wooden battens. Since it is now uneconomic to build thick load-bearing walls, cavity construction is used. In this method of construction the wall consists of two skins so that although the outer skin may be saturated the inner skin will remain quite dry if properly constructed.

An alternative technique is to construct the wall as an impermeable membrane as in glass curtain walling. Here the aim is to prevent

LOAD-BEARING CONSTRUCTION · 27

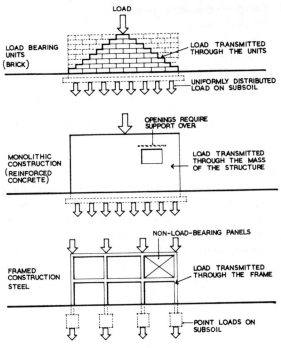

Diagram 15   Types of wall construction

completely the absorption of water. This construction requires very careful detailing, particularly at the joints which are the weak link in the constructional chain. Also, since the rainwater runs down the façade, care must be taken to form gutters at the base of the wall to take the water away.

All walling must be sufficiently strong to support the dead loading and superimposed loading which will be placed upon it. These loads, which are transmitted directly down to the foundations, are discussed in more detail on page 30.

The wall must also be so braced as to withstand the stresses caused by wind loading. Wind load is calculated on an average basis (N/m^2) over the surface of the wall depending upon the degree of exposure.

Insulation
When considerations of structural stability and weatherproofing have been satisfied, it is essential that the wall provides adequate thermal insulation, sound insulation and fire-resistance.

28 · BUILDING TECHNIQUES

The Building Regulations specify the minimum standard of insulation required for dwellings, and this is done by limiting the amount of heat that is allowed to be lost through the structure. This heat loss is indicated by a thermal transmittance coefficient known as the U value. This is the quantity of heat measured in watts which will flow from the inside to the outside of a construction through one square metre of the construction depending on the difference in temperature between the inside and outside environment. This temperature difference is measured per degree celsius. Thus the unit of measurement is expressed W/m^2 degC. A structure that is well insulated against heat loss will have a low U value, indicating that heat passes through the construction relatively slowly while construction that allows the heat to escape quickly will have a relatively high U value. Thus the various recommendations regarding minimum standards of insulation always specify maximum U values. The Building Regulations give a maximum U value for external walls of $1 \cdot 0$ W/m^2 degC. Diagram 16 compares various constructions.

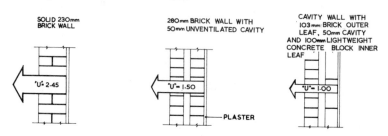

Diagram 16   Heat loss through walls

Previously the stipulated U values for dwellings have been concerned only with the solid part of the construction. That is to say, they did not take into account any heat lost through the window openings. In order to help to conserve fuel by restricting total heat losses in housing, the government through the 1974 second Amendment to the Building Regulations have introduced a requirement that the *average* U value of all the elements in the perimeter walls of dwellings shall not exceed $1 \cdot 8$ W/m^2 degC. The U value of a single glazed window is taken as $5 \cdot 7$ W/m^2 degC and a double-glazed window $2 \cdot 8$ W/m^2 degC. This means, in effect, that single-glazed window openings are limited to less than 20 per cent of the total wall area. This area of glazing can of course be increased by the use of double-glazing techniques. This provision is not a particularly successful means of full conservation since it does not take into account the orientation of the window. A south-facing window, for instance, fitted with heavy curtains can be an effective source of heat gain in autumn, winter and spring when the sun shines. The curtains are closed to retain the heat at night.

Since the density of a material is important in relation to its insulating properties, the density of the 100mm block indicated in diagram 16 should not exceed 800 kg/m^3.

The technique of filling the cavity between the two leaves of a wall of a house with waterproof insulation in order to increase the overall insulation value is now well established. The two types of material commonly used are waterproofed mineral fibre or a urea formaldehyde. There are various precautions that should be taken where cavity fill is used and so it is advisable to consult firms who have obtained an Agrément certificate for their particular technique. Cavity fill should not be used in conditions of severe exposure, but where conditions are appropriate very good standards of insulation can be achieved.

Sound reduction between rooms of different uses in buildings is important, particularly in the case of party walls in terraced houses or flats. The sound reduction value of a building element, such as a wall, partition or window, is denoted by a value known as a decibel (dB).

A reduction of 50 dB between rooms means that normal speech or radio sounds should not cause annoyance, but it is preferable to aim for as much improvement as possible.

For solid construction, sound insulation values increase with the mass of the structure; this is illustrated by the fact that the massively constructed load-bearing walls of old buildings were very adequate sound barriers while, on the other hand, the thin lightweight partitions adopted as an economic necessity in modern buildings are inadequate unless special measures are taken in the detailing.

It is difficult to measure accurately the total effect of sound reduction through an element since the building materials react differently to sound of varying frequency. Brickwork, for instance, prevents the transmission of low-frequency noise better than high-frequency noise.

The Building Regulations include a sound insulation requirement for certain party walls. In this connection it is important that account should be taken of adjoining structure, in that it is necessary to ensure that any party wall is adequately tied to the flank wall (usually at the front and rear of the building) or extends for a suitable distance beyond it.

For a party wall between houses, the Regulations stipulate that where solid construction is used it must be of sufficient thickness to give an average mass of construction of 415 kg/m^2 over a portion of the wall measuring 1 metre square.

Fire resistance

The further requirement of a wall is that it should resist the spread of an outbreak of fire for a specified period of time. This period is measured in minutes or hours and is termed the 'notional period of fire resistance'. The size, position and use of a building determines how long the period must be, and the Building Regulations set out the requirements. For example, the external walls of a two-storey house must have a minimum period of fire resistance of half an hour, while the external walls of a large multi-storey warehouse may have to have fire resistance of up to four hours.

Foundations

The weight of a building is transmitted to the ground through foundations. A foundation is the base of the wall or column which is extended to spread the load over a sufficient area of ground to keep the loads from the building and the resistance provided by the sub-soil in equilibrium. The principle is illustrated in diagram 17.

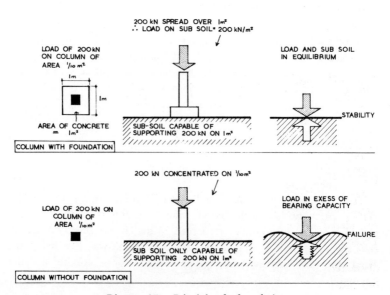

Diagram 17   Principle of a foundation

The dead load of a building is the weight of all permanent construction including floors and all the walls and partitions. Weights of materials used in calculations may be found in B.S. 648. The imposed loading is the weight of equipment, furniture and people using the building. To make calculation easier, the loading is averaged out for each occupancy. British Standards Code of Practice (B.S.C.P.) 3 chapter V specifies the loading allowance in kN/m^2 of floor area to be made as, for example: for houses, 1·5 kN/m^2; for general offices, 2·5 kN/m^2; for school classrooms, 3·0 kN/m^2; for shops, and for assembly halls, 4·0 kN/m^2. The foundations of any building are required to be designed to sustain safely both dead and imposed loads, so as not to cause excessive settlement and to be carried down to an adequate load bearing strata.

Settlement
Almost all buildings will settle into the ground after being built but the amount depends upon the loading and type of sub-soil. The settlement in domestic building will normally be very slight and will not be a problem, but on tall buildings which continue to settle over a number of years special constructional details would have to be incorporated to accommodate the movement. The foundations must also penetrate below the zone where shrinkage and swelling of the sub-soil takes place due to seasonal weather changes, particularly in the case of clay. Tree roots adjacent to the building can also cause considerable movement in clay soil several years after the trees have been cut down. The foundation depth must also be below the zone at which damage from frost may be expected. Trouble from this latter cause is most likely in chalk and fine sandy soil.

Types of foundations
Foundations may be continuous as in the case of wide or deep concrete strip foundations or they may be designed as a series of underground concrete columns known as 'piles' which are linked by a system of reinforced concrete beams at ground level.

The walls are built up off the foundation strip or beam and the ground floor of the building constructed at the correct level above ground. Alternatively, if the site is flat and the building is of a light frame and panel construction, the ground floor and foundations may be considered as one unit and constructed in the form of concrete raft. These alternative forms are compared in diagram 18.

Characteristics of sub-soils
Information should always be obtained on the type of load-bearing strata (sub-soil) likely to be found on the site of the proposed building,

32 · BUILDING TECHNIQUES

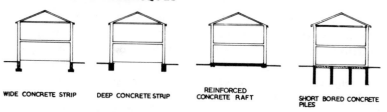

Diagram 18   Comparison of foundation types suitable for domestic building

before working drawings are started, so that the correct foundation can be chosen and if necessary the costs of alternative types can be compared. An intelligent assessment of the likely sub-soil conditions can be made from information obtained from adjacent building owners, District Surveyors and local builders, and reference to a geological map of the area will provide a general background. Following this, trial holes should be dug at relevant points on the site depending on the size and shape of the building. The trial holes should be dug to a depth below which the first load-bearing strata is encountered to ensure that this is not undermined by poor ground. Holes 3 to 4 m deep would be adequate for most single- and two-storey buildings. Details of types and depths of sub-soil encountered in the trial holes should be recorded, as to colour and consistency. The depth of water should be noted and, in particular, whether this is the natural water level, known as the 'water table', or whether it is entering at a higher level due to natural drainage. The drainage potential of the site can also be indicated by general topography and the warning given by hollows, soft conditions or marshy ground. Inspection of adjacent buildings for settlement cracks will also give an indication of difficult sub-soil conditions. The trial hole positions should be recorded on the working drawings for reference. For lightly loaded structure such as domestic building and single- and two-storey school or office buildings, further specialist site investigation involving expensive boring apparatus and laboratory testing will in almost all cases be unnecessary. Diagram 19 illustrates a specimen page from a report on a hypothetical site investigation.

Load-bearing strata can be divided into four groups: (1) rock; (2) non-cohesive soils such as gravel and sand; (3) cohesive soils such as clay; (4) peat and shifting sand. Rock in the form of limestones, sandstones, shale and hard solid chalk will take heavy loads but water may not be able to drain away and the top-soil may be waterlogged. Note, however, that if chalk is soft or badly weathered or the rock is thinly bedded the safe bearing capacity is very much reduced.

Dry compact gravel, or gravel sand sub-soil, provide adequate

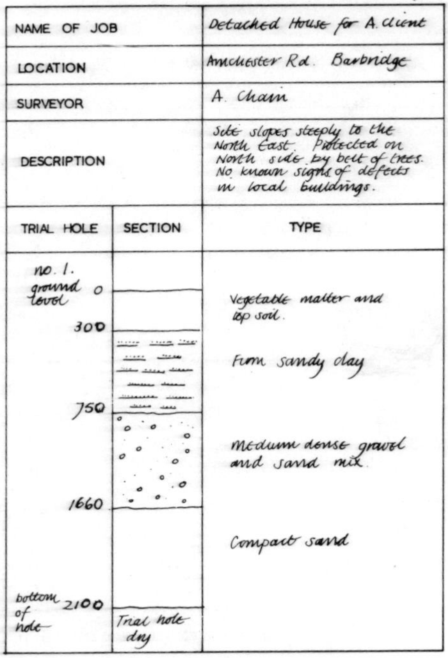

LOAD-BEARING CONSTRUCTION · 33

NAME OF JOB	Detached House for A Client
LOCATION	Anchester Rd. Banbridge
SURVEYOR	A. Chain
DESCRIPTION	Site slopes steeply to the North East. Protected on North side by belt of trees. No known signs of defects in local buildings.

TRIAL HOLE	SECTION	TYPE
no. 1. ground level 0 — 300		Vegetable matter and top soil.
750		Firm sandy clay
1660		Medium dense gravel and sand mix.
bottom of hole 2100	Trial hole dry	Compact sand

SPECIMEN PAGE OF SITE INVESTIGATION REPORT

Diagram 19   Site investigation

foundation, and are good to build on. But if the water table is such that the gravel is submerged in water, then the permissible bearing pressure is reduced by half.

Sand when damp, compacted and uniform holds together reasonably well and is a satisfactory foundation particularly where the sand can be retained by metal-sheet piling driven down to firmer soil below.

Clay is difficult to build on since the strength of different types varies considerably and clay generally is subject to long-term consolidation settlement. When clay absorbs water near the surface of the ground it swells and later drying out causes shrinkage, so there is always movement in the top two or three feet of clay soils.

Peat, and loose waterlogged sand are very poor sub-soils so that specialist advice is required before building, since large and unpredictable settlement occurs in these materials.

Safe bearing capacity

The maximum allowable bearing pressure in kN/m^2 is given for various types of sub-soil in Table A.

BUILDING TECHNIQUES

Table A

	Type	Allowable bearing pressure in kN/m^2	Remarks
Rocks	Limestone, and hard sandstone	1000 to 4000	Requires pneumatic pick for removal
Non-cohesive soils	Dry compact gravel	400	Can be excavated by hand or mechanized pick
	Compact uniform sand	300	Easily excavated
Cohesive soils	Stiff dry clay	300	Removed by mechanized spade
	Firm sandy clay	200	Readily excavated by hand spade
	Soft sandy clay	100	Easily moulded and readily excavated
	Very soft sandy clay or silt	50	Squeezes out when compressed

The figures allow a factor of safety based on experience of soil behaviour.

Although the total weight of a building, taking into account both dead and superimposed loading, will vary in specific cases the average weight on the foundations can be approximated for the more usual building types as shown in Table B.

From these two tables, the area of the concrete base required in a foundation can be calculated. By comparing the support obtained by

Table B

Type	Approx. load per linear metre run of foundations (kN)
Domestic single-storey dwelling of load-bearing brickwork	20
Domestic two-storey dwelling of load-bearing brickwork	30
Three-storey flats of load-bearing brick construction	120
Single-storey steel or concrete framed buildings	50 to 100 kN on columns

the safe-bearing capacities in Table A with the areas of the concrete-strip foundations shown in diagram 20, it will be seen that there is an additional factor of safety on most domestic work.

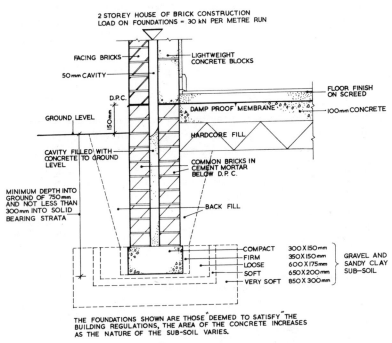

Diagram 20 Wide-strip concrete foundation

Wide strip concrete foundations

Traditional continuous wide strip unreinforced concrete foundations can thus be used for buildings of normal loading up to three stories in height, provided that the sub-soil is firm and compact and that there is no wide variation in the type of sub-soil over the loaded area. The Building Regulations require that the strip shall not be less thick than its projection beyond the base of the wall and in no case less than 150 mm. The width of the strip will vary according to the nature of the sub-soil as shown in diagram 20.

In practical terms it will often be found that the concrete will project say 150 mm beyond the wall in any case, in order to allow a foothold for the bricklayer.

The concrete strip should consist of a mix of one 50 kg bag of cement to 0·1 m³ of fine aggregate (sand) and 0·2 m³ of coarse aggregate (broken stone or gravel) to B.S. 882. Where the foundations are stepped, the thickness of the benching must be a minimum of 300 mm at the change of level. It may be less expensive, depending on the sub-soil variation, to excavate to a greater depth using mechanical equipment rather than to form steps in the concrete. It is good practice to put a layer of building paper in the trench before the concrete is poured, to prevent the water in the mix running away, with a consequent delay in the hydration (setting) of the cement.

Deep strip concrete foundations
In parts of the Midlands and the South East of England, the sub-soil is likely to be the particular type of clay which will be subject to excessive seasonal movements due to changes in the water content of the sub-soil. This clay is thus 'shrinkable' and a continuous, narrow, deep, concrete strip foundation should be used, which relies on the extra depth to withstand temporary lack of support due to the shrinking of the sub-soil at the base. A typical deep-strip foundation is illustrated in diagram 21. This type of foundation known also as 'trench-fill' can be used for most house foundations. The width and depth should be agreed with the Local Authority surveyor. The width will probably be between 450 mm and 600 mm. Ready-mixed concrete

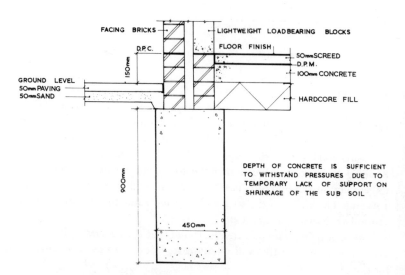

Diagram 21   Deep-strip foundation

can be poured easily into the trench and this method should show savings on labour costs against the wide strip foundation.

Short concrete pile foundations

An alternative method of providing safe foundations on shrinkable clay is to bore holes at given distances along the lines of the walls and to fill these with concrete to give a series of short underground columns or piles. The holes are bored by means of an auger mounted on the back of a tractor. This system is no more expensive than the deep strip, and it does not produce so much spoil. This helps towards a cleaner site, and the work can proceed when the weather is not suitable for trenching. It is, however, not suitable for places where large stones are found bedded in the clay, or where there are lots of tree roots. The load which is carried by the concrete beams must be carefully worked out so that the reinforcement can be calculated. The piles are placed at strategic positions in relation to partition walls and openings in the outer walls. Special reinforcement may be required if door openings occur near the corners of the building. This type of foundation is shown in diagram 22.

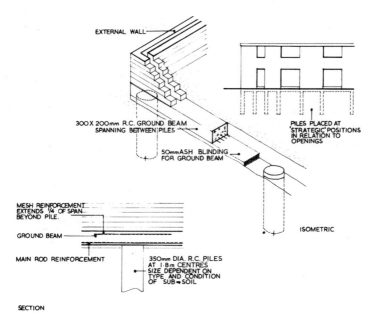

Diagram 22   Short-bored pile foundations

Concrete raft

A concrete raft foundation is used in conditions where the first 400 to 600 mm of sub-soil forms a crust which is stiffer than the underlying material. To build a raft under these conditions is often less expensive than penetrating the crust with foundation strips. It will be seen, however, that a decision whether or not to use a raft should be left to the expert with a detailed knowledge of the particular site conditions.

The raft will have mild steel reinforcing mesh at the top and bottom of the slab, as shown in diagram 23, and it is necessary to provide 1 m of impermeable surface around the perimeter of the raft as edge protection, as shown.

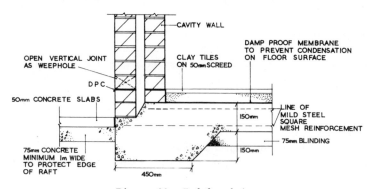

Diagram 23   Raft foundation

'Made-up' ground

In areas where suitable building sites are difficult to find, it may be necessary to build on 'made-up' ground or 'filling'. The 'fill' should have been made from brick rubble, ashes, slag, clinker or quarry waste, properly consolidated. Provided that the 'fill' is of carefully selected material and that up-to-date methods of compaction have been used, satisfactory support can be assumed, after a few months. If, however, a poor-quality 'fill' has been used, many years should elapse to allow subsidence to cease before contemplating using the site. In any case, it is always best to dig to the full depth of the 'fill' to check and to carry out soil mechanics tests to determine the degree of consolidation.

Reinforced concrete foundations

On good and reliable 'fill', traditional strip foundations may be used, provided that they are reinforced by mild steel rods or mesh. Transverse reinforcement of 12 mm bars at 150 mm centres with 10 mm diameter rods placed horizontally at 300 mm centres would be a typical reinforcement for a two-storey house. But, of course the reinforcement is always chosen after a calculation has been made based on a knowledge of the loading and sub-soil conditions. The reinforcement is laid with 40 mm cover from the bottom of the slab and it is a good plan to make sure that the base for the reinforced foundation is true and level by first laying a 'blinding' course of 75 mm weak concrete.

For heavier constructions, reinforced concrete piles and beams will be necessary. Particular care should be taken where the building is placed partly on 'made-up' ground and partly on natural gound. Here reinforced concrete piles should certainly be used to take the loads to a firm bearing on the natural ground below the 'fill'.

Concrete in foundations

Materials

Concrete consists of matrix (Portland cement), coarse aggregate (broken stone or gravel), fine aggregate (sand) and water.

There are two types of Portland cement used generally in concrete work: ordinary Portland cement and rapid-hardening Portland cement, both to B.S. 12.

The rapid-hardening cement has been more finely ground with some adjustment in composition, so that it develops its ultimate strength more quickly.

A third type used in special circumstances is sulphate-resistant Portland cement. This is a material formulated to give better resistance to attack by sulphate salts in the sub-soil, as discussed on page 42.

There are also other types of cement including high-alumina cement which has the property of gaining a high strength very quickly. The Building Regulations place restrictions on this type of cement due to recent experience of structural failure in buildings, mainly schools, which have been attributed to this material and thus it is no longer recommended at the present time for structural use.

Aggregate is defined as inert stone material ranging from sand (to pass a 5 mm sieve) to coarse material (to pass sieves having openings 19 mm or 37·5 mm square). The choice of the smaller or larger coarse aggregate depends upon the purpose for which the concrete is to be used. The whole mass of the aggregate must be very well proportioned within given limits. It must be clean and free from organic material, porous or weak particles or injurious chemicals. B.S. 882, 1201 is the Standard governing aggregate.

Water must be clean and free from injurious chemicals. It is now usual to expect that mains water will be used for building work.

Mixing

The aim is to produce the most economical mix for whichever purpose the concrete is to be used. A good-quality mix therefore must always be sufficiently durable and of adequate strength for the job in hand. The durability of the concrete is a measure of its resistance to weathering and frost action and also its resistance to chemicals in the soil or in the atmosphere.

Frost action is the most prevalent cause of damage, due to the absorption of water which on freezing expands, shatters the concrete at the surface and so makes further absorption, freezing, expansion and eventual disintegration possible.

The susceptibility of concrete to frost damage is associated directly with excessive water content at the time of mixing; but even when the right amount of water is used, there is still risk that the concrete mix will be damaged by frost before the setting action takes place, and so precautions must be taken when concreting in frosty weather. Concreting should therefore not be carried out with a temperature below $5°C$, and when frost is likely, the concrete should be well protected by sacking, sheets of plastic, or a covering of sand. The danger of frost attack is more likely, since cold weather delays the setting time of concrete. It is possible, however, to add a small amount of calcium chloride ($CaCl$), but not more than 1 kg to a bag of cement, to the mixing water to increase the rate of hardening of the cement. However this must be used with caution since $CaCl$ is a corrosive which may attack the steel reinforcement. The strength of the concrete is influenced by the water : cement ratio, the greater this is the weaker the concrete mix. Thus the concrete should be made with the lowest cement : water ratio possible compatible with satisfactory placing and compaction of the concrete. This is termed 'workability'.

Workability is increased by careful grading and selection of aggregates but unfortunately the easiest and cheapest way to achieve workability is to increase the amount of water, thus impairing the strength and durability of the mix.

The concrete in foundations will use either 20 mm maximum aggregate or 40 mm maximum aggregate. The 20 mm aggregate will give a stronger, denser concrete and will be suitable for concrete foundations to piers and reinforced strip foundations, while the 40 mm aggregate will be suitable for normal unreinforced concrete foundations.

The best way to ensure a good-quality concrete is to weigh the materials before mixing; this is known as 'weigh-batching'. When the aggregate is specified by dry weight, the mix is known as 'Standard' or 'Prescribed'. The weight of cement and any aggregates (in kg) to produce approximately 1 cubic metre of fully compacted concrete are given in B.S.C.P. 110 to produce several numbered grades of concrete for various purposes. Weigh-batching equipment is more economical on large contracts, but on smaller jobs the proportioning of the materials by volume may still be used, the materials being measured in gauge boxes. When the concrete is specified by volume the term 'nominal mix' is used and such mixes are quoted in B.S.C.P. 114. The proportions are $1:1:2$; $1:1\frac{1}{2}:3$; and $1:2:4$ of cement, fine aggregate and coarse aggregate according to the needs of the job. For complex structural design the Structural Engineer will specify the concrete by stating the performance required. When this procedure is followed the mixes are known as 'designed mixes' and the specification should include the type of cement, the maximum size of aggregate, the minimum amount of cement to be used and the required strength. The strength is usually checked by crushing test cubes of the concrete say at 28 days after pouring. After the concrete is proportioned, it will be mixed by mechanical mixer. Water should be added to the mixer in the form of a fine spray; the mixing should continue generally for 2 minutes. Coarse aggregate should be added first to prevent clogging and the mixer should be cleaned out thoroughly at the end of the day. Larger machines have divisions to act as gauge boxes.

The mixer should be as close as possible to the *in situ* position of the concrete. The most usual method of transporting concrete on the site is by barrow or skip. Excessive jolting must be avoided for this will cause the mix to separate out. If this happens, the liquid cement slurry will come to the surface and thus the effectiveness of the careful proportioning will be nullified. Concrete can be transported on the site by chute, but it must not be dropped from a height since this also will cause separation of the mix.

Approximately half the total *in situ* concrete used is now supplied 'ready-mixed' for delivery by approved suppliers by road. By this method the concrete is mixed under controlled conditions by the supplier and delivered at specific times as required by the building contractor.

Concrete must be placed as soon as possible after mixing and it should be thoroughly compacted, usually by a mechanical 'vibrator'. New concrete is called 'green concrete' and it must be allowed to set slowly, in order to develop its full strength. The protection of the concrete during the setting process is called 'curing' and this may be done by damp sacking, damp sand or plastic sheeting. Protection should continue for at least three days.

Sulphate attack
Foundation concrete must resist attack by sulphates in the soil. The salts which have a destructive action on concrete are the sulphates of calcium, magnesium and sodium. They occur in crystalline form in clays, particularly in the London and Oxford areas. The crystals dissolve into the ground water and form an alkaline solution which attacks the cement. In addition to the naturally occurring sulphate salts, it must be remembered that certain industrial wastes contain sulphates in solution which may find a way into the sub-soil. The severity of attack is increased by movements of ground water caused by drainage or fluctuations in the level of the water table.

Concrete which is subjected to water pressure from one side is more vulnerable, and when this situation is combined with free evaporation on the other side, the damage will be more severe. Thus deep strip or raft foundations are most vulnerable, while the risk of damage to traditional wide strip foundation is not so great.

The resistance of the concrete to attack depends on its quality and a well graded mix with ordinary Portland cement will normally be satisfactory, provided the concentration of sulphate salts is not excessive. Where the concentration is higher, the use of high alumina or super-sulphated cements will give greater protection.

Brickwork materials
Bricks are made of clay and shale. Southern England particularly has many areas of excellent clay suitable for brickmaking. There are usually varying quantities of sand, iron oxide and other substances already in the clay, and the variations in the proportion of these substances give a wide range of texture, colour, etc., which is one of the best characteristics of brickwork. Bricks can also be made with silica sand and lime. After mixing and moulding, they are steam-cured in an oven under pressure. This is known as 'autoclaving'. The resulting brick is rather like limestone, it hardens with time. They are strong, have an even texture, light colour and are termed calcium silicate bricks.

Bricks are also made of concrete. Coloured sands are mixed with cement and moulded to standard brick sizes. They are heavier than ordinary bricks and may crack on account of shrinkage, but they can be made in a pleasant range of colours.

Classification of bricks
Bricks are classified in many ways and the following is a brief guide:

Classification by use
Commons: Common bricks are comparatively inexpensive and are manufactured without regard for appearance. They are used in work below ground and as a backing material. Commons are classified to take a given loading depending on their position in the building.

Facings: The appearance of facing bricks is of primary importance and they are manufactured in a wide variety of textures and finishes. Facings must be carefully laid and afterwards protected so that the work remains undamaged. Nowadays, some facings are delivered pre-packed in small stacks, bound by wire and protected by timber pads.

Engineering: These are made to carry heavy loads and their appearance is not of primary importance, but because of the care needed in the choice of clay and during the manufacturing process, their appearance is often very acceptable.

Classification by finish
Sand-faced: One face and two ends are textured by the application of sand to the mould during manufacture.

Rustic: This is deep machine-applied texture in a variety of patterns.

Smooth or plastic: This is not an applied finish but the colour and quality of the clay makes a brick with smooth and rather shiny surface which is satisfactory for certain positions.

Facing bricks within the above range may be multicoloured and this will be indicated by the classification, e.g. multicoloured sand-faced facings. The colour of the brick may also be used for further identification, e.g. Staffordshire blue engineering.

Classification by manufacture
Wire cut: These are bricks manufactured from a ribbon of clay extruded in a continuous strip through dies giving the correct length and breadth (width) of the brick. This continuous ribbon of clay is then cut by wires to obtain the brick thickness (depth). One-third of all bricks are made this way. Wire-cut bricks may be commons or facings. They are always without a frog or recessed panel and may often have perforations of slits or holes formed to reduce the weight of the brick as part of the extrusion process.

44 · BUILDING TECHNIQUES

Pressed bricks: The process of manufacture for pressed bricks depends upon the consistency of the clay which is used. Stiff plastic or semi-dry pressed bricks have a characteristic frog on one or both sides. They are produced by mechanical process, the clay being confined in steel moulds.

When the brick unit has been produced by one of the methods previously described, the clay must be dried before burning. Drying should be carried out as slowly as possible to prevent subsequent shrinkage and deformation. After drying, the bricks are placed in a kiln. The tunnel kiln enables the bricks to be dried, burned and cooled in one continuous operation.

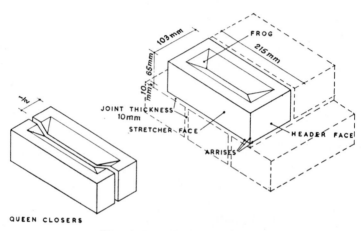

Diagram 24   'Work size' of a brick

The 'actual' or more correctly termed 'work' size of a brick conforming to British Standards is: length 215 mm, breadth 102·5 mm, and depth 65 mm. The breadth is usually rounded up to 103 mm for drawing purposes. Some account must be taken of the joint when detailing (and, of course, laying) brickwork, a 10 mm joint is usually assumed and this gives a British Standards 'format' size of: length 225 mm, breadth 112·5 mm, depth 75 mm. The 'format' size is also termed 'designated' size. Certain manufacturers also make bricks to 'modular' size's to relate to the recommended or preferred dimensions of 300 mm and 100 mm for general dimensional cordination in design. The 'format sizes' of the metric modular brick range are 300 x 100 x 100 mm; 300 x 100 x 75 mm; 200 x 100 x 100 mm; and 200 x 100 x 75 mm.

To take into account possible deformation of the brick during the burning process, B.S. 392 sets dimensional tolerances. For example, the tolerance over the length of 24 bricks when laid is ±75 mm with corresponding figures for breadth and depth.

Apart from the usual rectangular shape, clay bricks are commonly made with splayed or rounded edges which give great scope to the designer. These 'Standard Specials' are described in B.S. 4729.

Mortar

Mortar is the material in which brickwork is bedded. It binds the units together into a homogeneous mass making it possible to lay the bricks level and true. Since the mortar will fill all the spaces between the bricks, the overall sound and thermal insulation qualities of the mass of the wall are improved. The volume of mortar is approximately 12 per cent of the total volume of the wall and is therefore an important element of the construction. Mortar consists of sand with lime or cement added and mixed with water.

The water for the mixing must always be clean, and since drinking-water supplies are almost always available on a building site, this water can be used and tests for purity and cleanliness are not necessary. Water should be used in sufficient quantity to make the mass workable, and no more.

The sand should be evenly graded to B.S. 1200 from fine dust-like particles up to grains about 4 mm diameter. The grading can be tested by checking the percentage of granules retained on wire-mesh sieves. The sand must be clean, and free from shale and harmful soluble salts. It should also be reasonably free from clay and loam which should not be more than 5 per cent by weight of the total. The use of sand loam produces a mortar which is easy to use but of an inferior strength and which will have too great a shrinkage on setting. Harmful soluble salts are found in sea sand and so this should be avoided, since the salts would produce efflorescence on the surface of the wall or maybe cause dampness. Because of this, sand for mortar is always specified to be from a sand pit or river bed to be clean and sharp. Since sand is very costly to transport it is usual for a local sand to be used by the contractor, and so the source should always be checked to see that the material is suitable. For mortar into which stone masonry is to be bedded, crushed stone can be used instead of sand.

The matrix will be either lime or cement or a mixture of the two. Both non-hydraulic and hydraulic lime can be used to make mortar; non-hydraulic lime hardens and sets very slowly by drying out and carbonation by the air, whereas hydraulic lime sets by a reaction with water in a similar way to cement. The use of a non-hydraulic lime will produce a slow setting mass, whereas the use of an hydraulic lime will produce a stronger, quicker setting mortar, although in fact hydraulic lime is comparatively little used.

The cement used in mortar should be to B.S. and should be delivered in the sealed bags of the manufacturer and properly stored in dry conditions. This is because cement sets by hydration due to its reaction with water. Thus cement that has been affected by dampness should not be used.

Mortar mix

A mortar mix should be chosen in order to obtain a material as nearly as possible of equivalent porosity and strength to that of the bricks with which it is to be used. It is now usual for mortar to be mixed mechanically.

Cement mortar should be used below damp-proof course level. A typical mix is 1 part cement to 3 parts sand and this is usually written:

$$1 : 3 \text{ cement sand.}$$

This is a strong mix and is suitable for use with engineering bricks. Since the setting action of cement mortar begins immediately on adding water, mixes not used within two hours should be discarded. The proportions of lime mortar depend upon the degree of exposure to be expected, and a 1 : 3 lime sand mix would be adequate in a sheltered position. Since, however, lime mortar would not withstand severe exposure, cement is often added to the mix. This is then known as gauged mortar. Although considerable experience is necessary before a decision on the correct mix is made, the following will be a guide:

1 : 3 : 12 cement lime sand — suitable for internal work only.
1 : 2 : 9 cement lime sand — an average mix for normal exposure.
1 : 1 : 6 cement lime sand — suitable for severe exposure.

Mortar with plasticizer

The addition of lime to a cement—sand mix will make it easier to use, but nowadays it is probable that the Contractor would prefer to use a proprietary plasticizer. This will increase the ability of the water to surround and 'wet' the cement, increasing its effectiveness. Therefore less cement can be used while still retaining the overall strength of the mix. It is possible, by adding the recommended amount of plasticizer to a 1 : 6 cement mortar, to give a strength equivalent to a 1 : 3 mix. In addition, plasticizer entrains air bubbles in the mix which makes the mortar easier to use and also more 'elastic' when set, which in turn reduces subsequent cracking. Note that the season of the year is important in determining the proportions of the mix. In general terms, the colder the weather, the stonger the mix must be. Cement is thus often added to shorten the setting time.

Laying bricks
Since the aim in bricklaying is to obtain a solid mass of masonry, the bed and side joints must be 'well flushed up'. In order to achieve this, bricks in good-quality work should be laid 'frog up'. If they are laid 'frog down', there will be a high proportion of void in the finished wall as illustrated in diagram 25.

Jointing and pointing
The technique of laying bricks in mortar is known as jointing. The technique of forming a face joint on the brickwork is known as pointing. The wall may be 'jointed and pointed' as the work proceeds or,

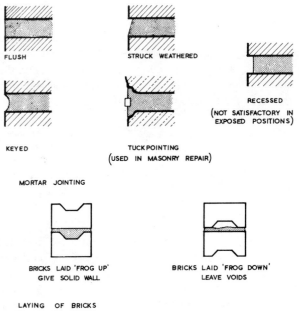

Diagram 25   Jointing in masonry

48 · BUILDING TECHNIQUES

alternatively, the pointing may be carried out as a separate process a little while after the wall is built. There is also a separate technique of scraping out the joints and repointing associated with repair work. Diagram 25 shows the profile of the joints in common use.

Where it is desired to emphasize the joints, it is probable that a recessed joint will be specified. When this is done it should be remembered that the lip at the base of the joint will collect water and so the weathered joint is to be preferred to the recessed joint shown in diagram 25.

Brick walls

The bonding of brickwork helps to distribute the loads from floor joists, beams, etc., evenly along the length of a wall and also gives an interesting pattern to the face. Many patterns are possible, but English and Flemish bonds are the most common. These are shown in diagrams 26 to 29.

Each row of bricks is called a course. In English bond alternate courses show the long or stretcher face of the bricks and the end or header face. To break the joint or to give the proper lap, part of a brick, called a queen closer, has to be placed near the angle or quoin. It is placed next to the first header.

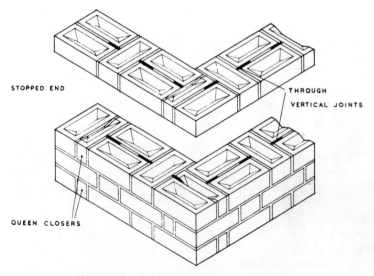

Diagram 26   Single-brick wall — Flemish bond

LOAD-BEARING CONSTRUCTION · 49

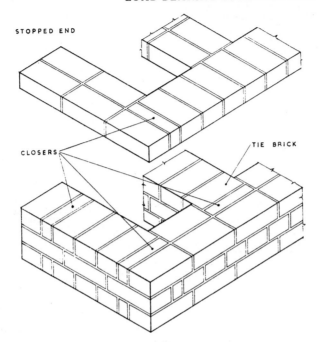

Diagram 27   Single-brick wall — English bond

In Flemish bond each course has header and stretcher alternately, but again a queen closer is necessary and this also is placed next to the first header. It will be found that this simple rule of placing a closer next to the first header in a course will give a proper bond. The stopped end shown in diagram 28 is an exception: the placing of a closer next the header at the stopped end would have made complications in the bond in the length of the wall, so a three-quarter-brick or bat is used.

With good mortars bonding does not much affect the strength of the wall, but in Flemish bond there are places in the thickness of the wall where the bricks do not overlap. These are shown in heavy lines in the diagrams.

It is essential that walls be bonded together at junctions, and this is usually done with a tie-brick every alternate course, as shown in diagram 27.

With an increase of thickness, more complex arrangements are necessary as shown in diagrams 28 and 29. In Flemish bond (diagram 28) half-bricks or bats have to be used in the middle of the wall.

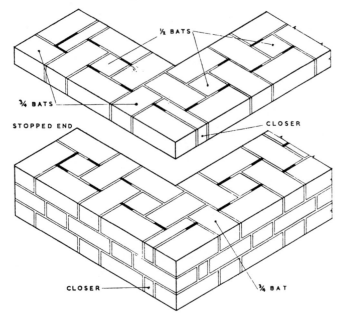

Diagram 28 One and a half-brick-thick wall — Flemish bond. Heavy lines indicate through vertical joints

It is important that the vertical joints of alternate courses line up. This lining up is called 'perpend'.

One brick thick solid walling, unless cement rendered, is not water-resistant in exposed positions and will in any case not provide the minimum standard of insulation required by the Building Regulations. It is thus standard practice that all habitable buildings are constructed in cavity walling. Solid walling is used to form the internal dividing walls in cross wall construction or for the external walls of farm buildings and sheds. One and a half brick thick walling may, in addition, be used below ground as base to cavity walling above.

Cross-walling

Cross-wall construction has become increasingly used for building terraced and semi-detached houses and for three- and four-storey blocks of flats.

Cross-walling has obvious economic advantages, since the end and party walls between dwelling units would in any case not have openings in them, and must act as adequate fire breaks and sound barriers as well as give lateral stability. Thus a simple unpierced load-bearing wall

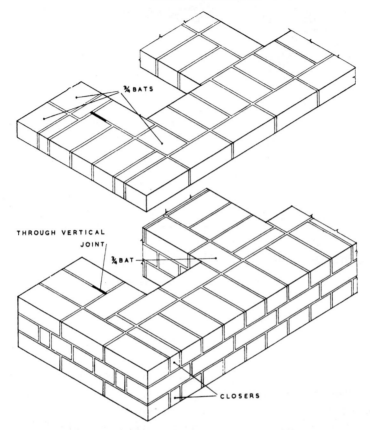

Diagram 29 One and a half-brick-thick wall — English bond. Showing quoin and junction

becomes the most efficient means of support. This means that the front and rear walls are relieved of any superimposed load and can become non-load-bearing panels between floors and roofs required only to keep out the weather and not needing elaborate foundations. Diagram 30 compares cross-wall construction with traditional load-bearing brickwork.

If the roof is flat, composite beams as shown in diagram 30 can be used for both floor and roof structure. Pitched roofs are generally designed to be supported on the front and rear walls of a building and cross-wall roofing therefore requires different solutions from the traditional purlin and rafter roofs. One method is to introduce beams or

52 · BUILDING TECHNIQUES

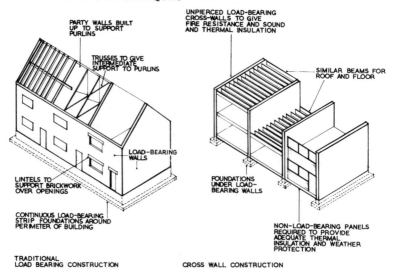

Diagram 30   Comparison of load-bearing and cross-wall construction

'trussed purlins' to support the common rafters, and round which the roof timbers are framed.

Arches and lintels

There are many ways of carrying the brickwork over openings. Concrete lintels as shown in diagram 31 are very strong. The ends should have a bearing of 150 mm on the wall. Small ones can be made on the ground and lifted into position, or they can be cast *in situ*. The traditional way was to form an arch, and diagrams 32 and 33 show two of the simplest types. The segmental arch needs a shaped piece of wood, called a centre, to be fixed temporarily for placing the bricks on until the mortar has set. No cutting of bricks is required for this arch. To set out this arch, the rise must be decided upon and then the centre of the radius can be found by drawing a line from the springing to the top of the rise and bisecting it. The centre of the radius is where this bisecting line meets the centre of the opening.

Soft bricks can be rough-axed to a wedge shape as voussoirs to form an arch, or a very neat appearance can be made by using very soft bricks, called rubbers, and shaping them to exact sizes, as shown in diagram 33, to form a camber arch. In this case the joints can be very thin. If the under edge or soffit of this arch is quite level, it will appear to drop in the centre. It is therefore necessary to give a small rise as

LOAD-BEARING CONSTRUCTION · 53

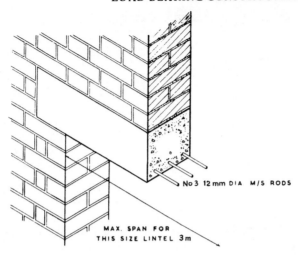

Diagram 31   Reinforced-concrete lintel

shown to avoid this optical illusion. For these arches the angle of the skewback is decided upon, usually 60 degrees; this line extended down will intersect the centre line of the opening and give the centre from which the radiating joints must be set out. Each arch brick or voussoir should be cut to a pattern or template.

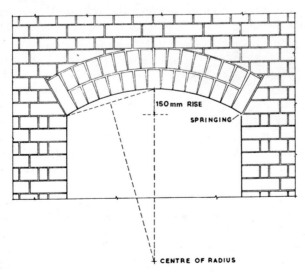

Diagram 32   Segmental arch of two half-brick rings

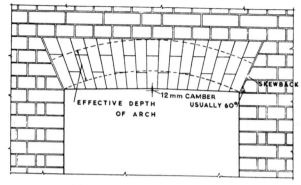

Diagram 33   Gauged camber arch

It is very necessary to make sure that horizontal surfaces in brick walls are well covered, to prevent rain lying and soaking in. Sills should be carefully designed and made of impervious material. A stone sill is shown in diagram 34; it is weathered to take the rain off quickly and has a throating under the projecting edge so that the rain drips clear of the brickwork below. The projection should be not less than 50 mm. As each end of the stone sill is built into the brickwork, a stooling is formed which stands up slightly and upon which the brickwork at the reveal is bedded. By this means rain is prevented from flowing over the end of the sill where the additional washing by rain will pattern the face of the wall.

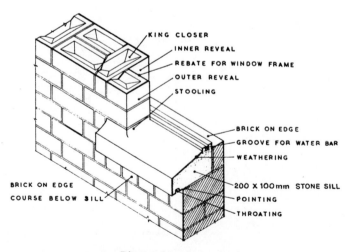

Diagram 34   Stone sill

LOAD-BEARING CONSTRUCTION · 55

Sometimes sills are set in after the rest of the wall is built. In any case it is advisable not to put the mortar under the sill until after the brickwork has set, otherwise any slight movement of the brickwork at the jambs may break a stone sill at the centre. Other materials are used for sills, some of which are shown and described later.

A stone coping is shown in diagram 38. It should have a good fall to the back and good throats under each projection. If the stone is not impervious, a bituminous felt damp-proof course should be formed below it. Cast stone, made of cement and stone aggregate, is often used. It shrinks slightly, so the joints in copings tend to open and a damp-proof course is particularly advisable in this case. Brick-on-edge copings, as shown in diagram 81, are traditional construction and are satisfactory in sheltered positions. The tile creasing, two courses of tiles in cement mortar, should be made of dense tiles.

Thresholds of openings should be of impervious material and should not interfere with the continuity of the damp-proof course. Stone thresholds were used mostly in the past, but brick ones, as shown in diagram 35, are now sometimes specified.

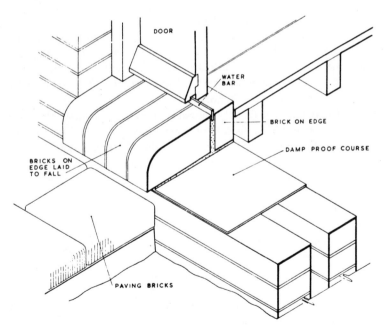

Diagram 35   Brick-on-edge threshold

Damp-proof courses

In order to prevent moisture from the wet sub-soil creeping up the wall by capillary attraction, it is necessary to insert a layer of impervious material known as a damp-proof course (d.p.c.). The d.p.c. must be placed sufficiently above the ground level to prevent the construction becoming wet by rain splashes. 150 mm above ground is considered to be the minimum; at the same time, however, the d.p.c. in the wall must be linked to the damp-proof membrane (d.p.m.) where the floor construction is solid, as shown in diagrams 20 and 21.

A d.p.c. must be impervious to moisture, durable and obtainable in relatively thin sheets and preferably pliable. A list of suitable materials is given below together with comments on each:

(1) Bituminous felt with fibre, hessian or asbestos base to B.S. 734. This is inexpensive and easy to lay, but liable to damage during the progress of the work. The bitumen may squeeze out under heavy loading, leaving only the base.
(2) Bituminous felt with a very thin core of lead, copper, aluminium or polythene sheet. These sheets are more expensive but overcome many of the disadvantages of non-core types.
(3) Copper sheets to B.S. 2870.
    The sheet normally used will be not thinner than 0·25 mm. This is expensive but reliable and adaptable to awkward shapes.
(4) Lead sheets to B.S. 1178, weighing not less than 19·5 kg/m^2. Lead is expensive but very reliable except where attacked by acids through cement mortar.
(5) Heavy-grade black polythene sheet having a thickness of not less than 0·46 mm. This is light, comparatively easy to lay, and very tough, though easily punctured by the edge of a trowel.
(6) Mastic asphalt to B.S. 1097 laid in two layers to finish 20 mm thick. This is very reliable as a d.p.c., but requires skilful operatives and is only economic where other asphalt work is being carried out at the same time. The asphalt should be kept back 13 mm from the face of the brickwork to allow for pointing — which will be quite a thick 'line' on the face of the building.
(7) Two courses of good-quality slates set in cement mortar and laid breaking joints. The slate is expensive but is not damaged by direct heat or direct pressure. It cannot, however, easily withstand differential settlement.

LOAD-BEARING CONSTRUCTION · 57

(8) Engineering bricks are expensive except where the bricks are made locally. They provide a strong, sound base for a building where a plinth is desirable.

(9) Damp proof courses manufactured from pitch polymer. They carry heavy loads, but are not covered by a British Standard and so those having Agrément certification should be chosen.

The sheet materials are available in normal wall widths and are laid with 150 mm laps at the joints. A folded (welted) joint is preferable with metal d.p.c.'s.

Damp-proof membranes
A horizontal d.p.m. is required to give protection to a large area and is normally sandwiched within the construction as in solid ground floors, and in basement construction. For solid floors, asphalt or bituminous compounds applied by a trowel or brush or spray are most satisfactory.

Special techniques are required where the construction is below ground level, as in basements or on sloping sites. The asphaltic and bituminous compounds are suitable in such cases as shown in diagram 36.

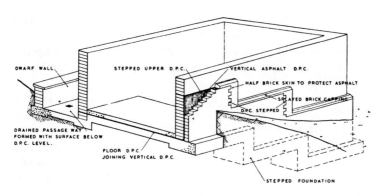

Diagram 36   Stepped foundations and vertical damp-proof courses

Rendering
In order to make a single brick wall more weather-resistant or to improve the appearance of old worn brickwork, it can be covered by a mixture of cement and sand or cement, lime and sand, known as 'rendering'.

For external rendering the purpose is usually to obtain a reasonably waterproof surface, so that cement and sand are chiefly used, mixed in the proportion of 1 : 3. Cement mixes tend to shrink as the cement sets, so it is important that the backing is very hard and rough to give good adhesion. Rough bricks are excellent, but smooth bricks must be keyed or hacked. For the same reason rendering will only adhere to concrete if it has a rough face, such as is obtained by exposing the aggregate by brushing before it is set. Lathing to take external rendering should be fixed very firmly.

Two coats of rendering are normally used for external work, the first a thicker coat to bring the surface up to a relatively true face. This should be well scratched for a key for the second coat. The second coat is thinner and enables a very true face to be obtained. Trowelling with a wood float or metal trowel is most commonly used in this country, but too much trowelling brings the cement to the surface, which will cause crazing. The Continental practice of throwing on each coat is good, as better adhesion is obtained. Trowelling is only necessary, then, to smooth the surface. Plasterers who have the knack can throw the plaster on with a flick of the trowel, but machines are available which will spray the plaster on to the surface in a suitable manner.

Smooth finishes for external rendering are not so satisfactory, as they can only be obtained by the trowelling referred to above. Rough textures obtained by brushing down, scrubbing when only partly set, or even the thrown-on plaster left unsmoothed are most satisfactory, although such surfaces will get dirty more quickly.

Cavity wall construction
The functioning of a cavity wall is shown in diagram 37. The outer skin must be kept clear of the inner skin. Metal ties enable each skin to help support the other, and these must be arranged with notches or bends so that water cannot run across on them and so that there is no horizontal surface for mortar droppings to lodge on. They are usually spaced 1 m apart along the wall and 6 courses apart vertically and staggered. Where it is necessary to bridge the cavity or join the two skins together, impervious flashings must be used to keep the inner skin dry as shown at the lintel and also at the side or reveal in diagram 38. The back of a parapet must also have a damp course and flashing as shown. Great care must be shown in these details and also in the supervision of the work, as the outer skin is very thin and water may, in a storm, actually run down its inner face in a stream and any bridge will cause a large damp

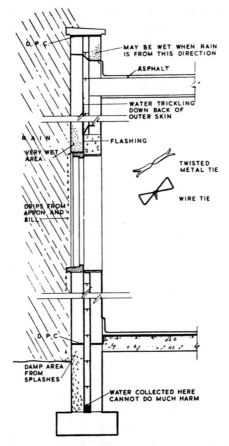

Diagram 37   Functioning of a cavity wall

patch to appear on the inner face. For this reason the cavity must be kept clear of mortar droppings as it is being built. Holes should be left temporarily at the base of the wall so that the bottom of the cavity can be cleared out frequently.

The cavity can go right down to the concrete foundations or it can be filled with concrete up to ground level, or a solid base can be built. To ventilate the space under a wood ground floor, airbricks and ducts have to be formed. These ducts can be lined with slate to avoid ventilating the cavity, as discussed in more detail on page 90 and shown in diagram 70.

Diagram 38 shows a concrete lintel which supports both skins. Other

60 · BUILDING TECHNIQUES

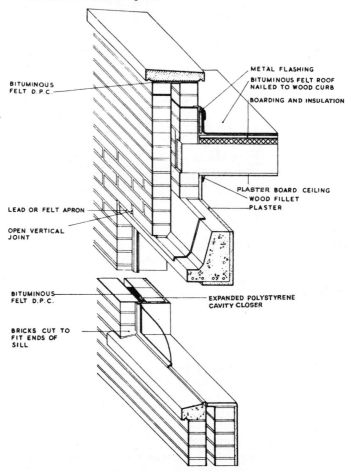

Diagram 38   Cavity wall – details at parapet and window opening

ways of spanning openings are shown in diagram 40 where the brickwork is reinforced with rods, and in diagram 41 where a metal angle is used. These also show a sill made of two courses of tiles set in mortar and two ways of treating the reveals, one where the cavity is closed and a rebate formed to take the frame, the other where the cavity is not closed but covered with the frame. This is suitable for protected positions.

Diagram 39 shows a metal sub-frame which is built in, and into which a standard metal window fits. This closes the cavity at the reveal, acts as sill and will support the outer skin over openings up to 1 m wide.

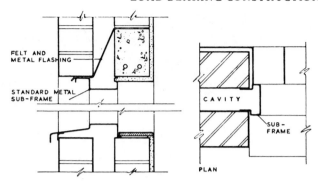

Diagram 39   Metal subframe – window opening in cavity wall

At the top the cavity should be closed. This can be done with a course of slates or as shown in diagram 41, which also shows a simple method of constructing the eaves of a pitched roof.

Fireplaces and flues

The Building Regulations refer to three classes of heat-producing appliances; they include:

(1) Class I, a solid-fuel (or oil-burning) appliance having an output rating not exceeding 45 kW.
(2) Class II, a gas appliance having an output not exceeding 45 kW.
(3) A 'high-rating' appliance, which can be solid-fuel, oil or gas burning having an output exceeding 45 kW. The 'high-rating' class are heavy-duty appliances and will normally be installed

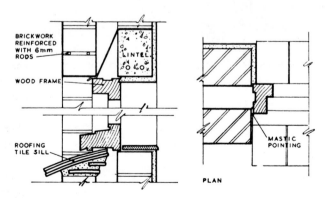

Diagram 40   Reinforced brick lintel in cavity wall

62 • BUILDING TECHNIQUES

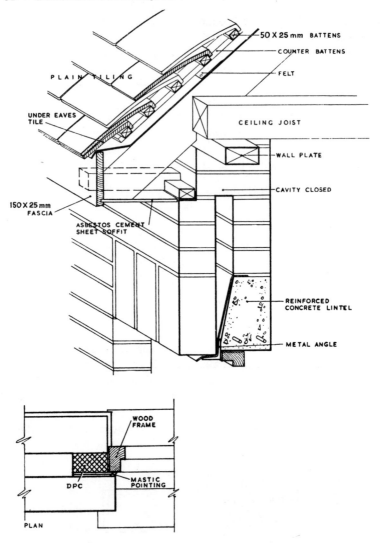

Diagram 41   Cavity wall — details at eaves and window opening

under the supervision of a Heating Engineer. Household fire appliances fall into Class I.

A firebrick back is used to line the fireplace. This is fire resistant, will not be damaged by heat, and when it gets very hot it radiates the heat into the room. It must be set on a concrete hearth min. 125 mm thick and coming at least 500 mm out into the room beyond the fire opening. The brickwork over the fireplace is gathered over in the manner of a funnel until a duct or flue usually 225 x 225 mm is left. This is carried up in the manner shown in diagram 42 and out of the roof, preferably near the ridge. It is convenient to collect the flues together to form a chimney breast.

It is now considered the best practice to form a lintel over the fireplace opening and leave only a narrow opening or throat 100 mm wide and 300 mm long for the smoke to go through, as shown in diagram 43. This produces a quicker draught at this point and prevents downdraught.

A hearth at ground-floor level can rest on a fender wall with

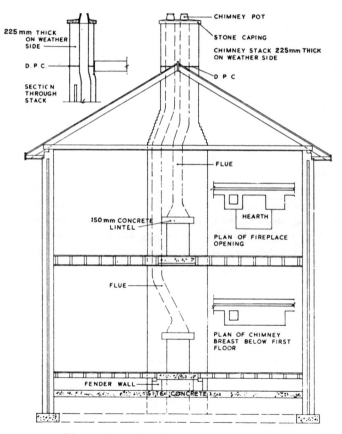

Diagram 42   Flues, chimney breast and stack

64 · BUILDING TECHNIQUES

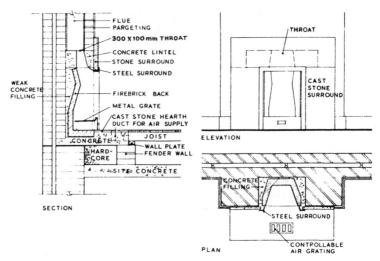

Diagram 43   Ground-floor fireplace

hardcore filling as in diagram 43. A hearth at an upper-floor level is best formed as a concrete cantilever reinforced with rods as shown in diagram 44. The edge against the wall must go 100 mm into the brickwork. Concrete hearths supported on boarding attached to the trimming joist are more common (diagram 45), but if the trimmer deflects slightly the hearth is likely to crack.

The finish of the fireplace to the room can be treated in many ways, but usually consists of a metal frame to the opening and a surround of tiles, wood, stone or brick, usually made into one piece which can be fixed back to the wall with metal cramps. The hearth can be tiled.

Flues must have at least 100 mm of brickwork round them and be rendered on the outside where any wood comes near. No wood may, of

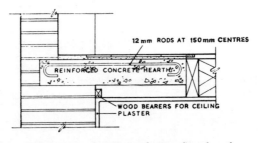

Diagram 44   Cantilevered upper-floor hearth

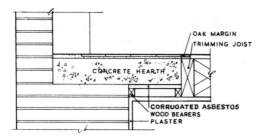

Diagram 45   Trimmed upper-floor hearth

course, go into this flue brickwork, or into the brickwork near the hearth closer than 250 mm to the fireplace or flue. Brick flues serving solid-fuel or oil-burning appliances must have a lining of fireclay, glazed clayware pipes or other suitable pipes called 'flue liners'. These liners should be rebated and socketed and jointed in cement mortar. A typical fireclay liner is shown in diagram 46.

It is important that the chimney stack be kept as dry as possible, so it should have a good covering, such as an impervious stone coping or tile creasing with a projection and throat, and a damp-proof course should be inserted at the level of the roof intersection (diagram 42). If the flues can have 225 mm of brickwork round them in the stack, particularly on the weather side, they will keep warmer and the fire will draw better.

Flue for domestic boiler
The construction of a flue to a domestic boiler of the type commonly used in connection with small-bore central-heating systems needs special care. Domestic boilers are designed to make efficient use of the fuel which they burn and this means that there is not so much heat left in the flue gases. Thus, the water vapour which is always produced as a byproduct of the combustion of the fuel condenses out. This did not happen with the ordinary open fire since the flue carried much more warm air as wasted heat up the chimney. The condensate absorbs other byproducts of the combustion to form a corrosive liquid which will attack the brickwork of the flue. Stains will appear on the surface and in very bad cases the flue will distort and crack. In order to prevent this, the flue should have an impervious liner and should also have the means for collecting the condensate which may run down the chimney. Diagram 46 illustrates the flue connection to a domestic boiler and shows a standard glazed fireclay flue liner.

66 · BUILDING TECHNIQUES

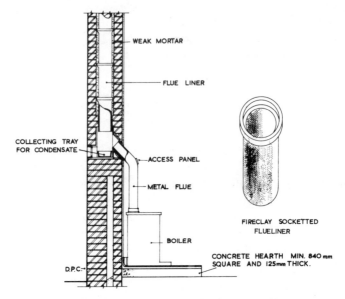

Diagram 46   Flue connection for domestic boilers

Gas fires need flues but they can have a smaller sectional area. Several types of proprietary concrete flue blocks are made, which will bond in with brickwork, and are usually 225 mm thick overall, thus avoiding the necessity for a projecting breast.

Stonemasonry
Stone has been a traditional material for building in Britain for hundreds of years. Because it could be quarried in large blocks, it was ideal for building the load-bearing walls and columns for castles and churches. Mediaeval masons understood the qualities of the material very well and by the intricacy of the carving and knowledge of the geometry of structure they produced many fine buildings. Stone was still the principal material used to face the many large country houses and public buildings in the cities during the seventeenth century and later.

With the advent of the steel frame, however, the great strength of stone was no longer required and in fact its weight became a disadvantage. Thus its use on major building projects has undergone a fundamental change and it is now usually required in thin slab form as facing material. The slabs are held back to the structure by non-ferrous cramps but it is advisable for the main weight of the facing to be taken by corbels or brackets designed as part of the main framework or by bonding courses as shown in diagram 53.

Rubble walling, either random or coursed, is often used in domestic work and for screen walls where the attractive colour texture and weathering qualities of stone are desired.

CHAPTER FOUR

Framed Construction

Multistorey buildings can be erected quickly and cheaply if a frame of steel or reinforced concrete members is used. This frame carries all the load: floors, floor loads, partitions and the outside walls. The walls can be quite thin, as they support only themselves, serving to keep out the weather and give insulation. These are sometimes called 'panel walls' or 'curtain walls' and their thinness saves floor area, which is usually valuable in high buildings; 150 mm to 225 mm thickness is possible. Load-bearing brick walls of old non-framed buildings such as warehouses were as much as (1200 mm) thick at the base when seven or eight storeys were built.

Structure steelwork
Steel is an iron with additional small amounts of sulphur, silicon, phosphorus, manganese and sometimes a little copper. Steel members of I-section are usual for the frame and they can be connected by riveting, or more commonly welding, and site bolting to angle cleats, as shown in diagram 47. Welding, for the type of joint shown in diagram 48, is widely used. The steel frame must be designed and detailed by a

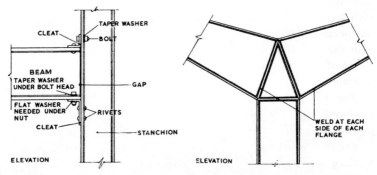

Diagram 47   Riveted and bolted steelwork connection (assumed pin-jointed)   Diagram 48   Welded steelwork connection (rigid)

FRAMED CONSTRUCTION · 69

Chartered Structural Engineer, who will calculate the possible loads on beams and stanchions and choose steel members of appropriate size, so that the stress in the material does not exceed the 'permissible stress' at any point. The allowable stress in bending, shear and direct force and maximum deflections are laid down. The latest revision of British Standards Specification 449 gives a standard of good practice.

The material used for structural steelwork will be steel (weldable) conforming to B.S.4360, which includes four tensile grades numbered 40, 43, 50 and 55 hbar (1 hbar = 10 N/mm^2). In addition to the main grades there are several subgrades in each range, and steel sections produced to a particular grade are marked with the appropriate designation, i.e. 43, followed by a letter indicating the subgrade.

A special type of low-alloy weathering steel which has been developed for use unprotected outdoors in the USA is now available in the UK and is known as 'Cor-Ten'. Initial rusting is, in fact, allowed to take place, but because of the special properties of the steel, continued exposure builds up a self-protective surface. The colour of the weathered steel varies from light brown to purple and the formation of the protective layer takes a considerable amount of time depending on the external conditions of exposure.

Structural sections are 'rolled' from white-hot steel ingots into H-beams, channels, T's and angles. Seamless hollow tubes, square, rectangular or circular in cross section are also available, and are now widely used for roof trusses, purlins and secondary structural members. They are marketed as Structural Hollow Sections, abbreviated S.H.S.

A comparatively small range of steel joists are produced to a slightly different internal profile to a Universal Section for use as simple beams. Diagram 49 shows a selection of the range of structural steel sections now available and gives the maximum and minimum dimensions for each section. The size quoted for any Universal Section is a nominal one because they are rolled in various thicknesses of web and flange but with the dimension between the flange plates constant, to make the design of connections easier. This means, however, that the overall size increases in some examples by as much as 100 mm on the overall depth of the section. Although tubes and rectangular sections are also produced in a variety of weights for each designated size, the outside dimensions are kept constant.

Each Universal Section is given a serial number and it is usual on working drawings to use the serial number of the section and also to specify the size of the section, depth first thus: 203 x 133 x 30 kg/m U.B.

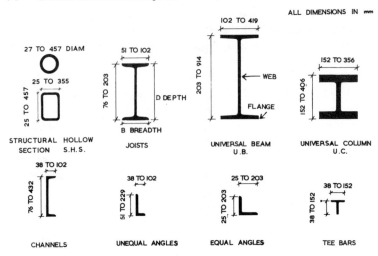

Diagram 49   Range of structural steel sections

In spite of the actual rigidity of connections, beams are often designed as if the ends had a free bearing on the stanchion, i.e. 'pin-jointed' (diagram 47). The whole of the beam is assumed to be free to deflect, and the load put on the stanchion is then vertical. In certain instances it is now more economical to design these joints as 'rigid', so that the stanchions help to stiffen the beams. The difference is shown in diagrams 50 and 51. In diagram 50, with assumed pin-jointed connections, each beam has an assumed effective span of the full distance between centres of stanchions, and the whole of this span can sag or deflect when the load is put on. This deflection is shown diagrammatically at A and C.

When the connections are rigid, as with welded steel frame, or reinforced-concrete frame, as in diagram 51, beams in line become one continuous beam, the parts next to the stanchions acting as double cantilevers and the parts of the beams between these acting as beams of much less span and so of much less deflection (B). Members can therefore be smaller, which in turn saves weight, and consequently load, on lower stanchions. In a tall building saving of dead weight of a frame can be important.

The transverse beams or cross-beams are also reduced in effective span if rigidly connected to the stanchions. This is shown at D (diagram 51). These beams are, however, trying to bend the stanchions inwards and outwards so the stanchions have to be strengthened.

FRAMED CONSTRUCTION · 71

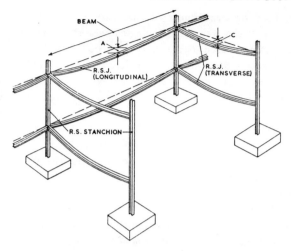

Diagram 50   Steel frame assumed pin-jointed (effect of loading exaggerated)

To make rigid connections with bolts and rivets is difficult. Welding simplifies rigid connections and has completely replaced riveting for workshop connections. Local additional strength can be obtained by welding on small pieces of steel wherever required. Most welding is done in the workshop since site welding requires careful supervision and is not generally recommended.

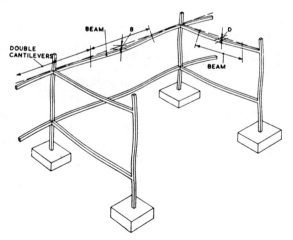

Diagram 51   Concrete frame with rigid connections (effect of loading exaggerated)

In an I-section beam most of the work is done by the top and bottom flanges, the web only serving to keep them the right distance apart. If welding is used, two T-section members can be used and held apart by spacer plates welded on where required. Welded steelwork needs even more complicated design if the maximum amount of steel is to be saved. Note that the welded connection is much neater than the riveted.

Diagram 52 shows a typical layout plan of a floor in a multistorey steel-framed building, showing spans which are common for concrete floor slabs and main steel beams. The edge beams only support the panel wall and the tie-beams make erection easy or give stability. Additional steelwork will be needed to support the landing, the lift and parts of the staircase.

In diagram 53, a traditional form of framed construction is shown. A single brick wall has thin slabs of stone fixed on its face, held mostly by cramps but resting at certain levels on a corbel course or thicker stone built into the core. The support of stones that come over windows is always a problem. Sometimes it is possible to carry them by special cramps at their edges, i.e. in the vertical joints. In diagram 53 they are shown carried on metal corbels set in the concrete casing.

Diagram 54 shows a more modern form of construction. Here a half-brick panel stands on the edge of the floor. Behind the brick panel and separated by a 50 mm airspace is a 100 mm lightweight concrete block insulating panel.

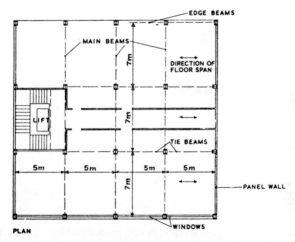

Diagram 52   Layout of steel frame for multistorey office building

FRAMED CONSTRUCTION · 73

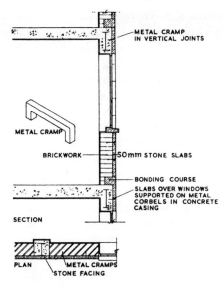

Diagram 53   Panel wall — brick with stone slab facing

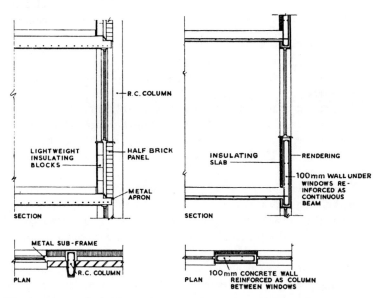

Diagram 54   Cavity-brick panel wall   Diagram 55   Reinforced-concrete wall

This panel is shown in conjunction with a concrete frame. Concrete frames are fully rigid. It is essential that skilled men be employed on shuttering and placing of the reinforcement and concrete and that there is constant and efficient supervision, and of course that the whole of the design be carried out by a competent engineer.

Concrete can be used for the external walls of a building as well as for floors and roofs. It is possible to design walls as little as 100 mm thick, reinforced so that the part of the wall below the windows acts as a continuous beam of considerable depth but very thin (diagram 55). Such a wall would need good insulation on the inside against heat loss.

It is not easy to get a satisfactory appearance with a concrete surface. The concrete should be of exactly the same aggregate and cement throughout the whole façade, and the water : cement ratio always the same. Aggregate should be chosen carefully from the point of view of appearance and trial panels made. The aggregate can be exposed in several different ways, which will improve the weathering qualities of the concrete.

Timber framing

Timber has always been a traditional material for framing structures, demountable timber-framed buildings being commonplace for housing during the fifteenth and sixteenth centuries. The widespread use of modern techniques in the treatment of timber for protection against termite and fungus attack, and the introduction of new adhesives and manufacturing techniques for producing timber structures in new forms makes timber framing a very acceptable contemporary method of construction.

Timber can be used either as a complete wall frame or as a series of panels between the main structural members. Diagram 56 shows a wall-frame method of construction developed in Canada. The timber framework should be protected by both a moisture barrier and a vapour barrier.

The moisture barrier, which is a moisture-proof but not vapour-proof building paper, is fixed on the colder side of the construction between the outside cladding and the sheathing as a second line of defence against the penetration of rain through the brick (or timber) facing.

The vapour barrier is placed on the warmer side of the construction immediately behind the internal wall lining and its purpose is to

FRAMED CONSTRUCTION · 75

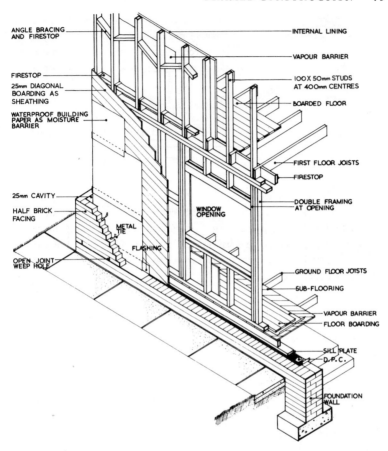

Diagram 56   Timber wall framing

exclude from the timber structure the moisture-laden air from inside the building. Aluminium foil is a satisfactory material for this purpose, and therefore foil-backed plasterboard is a very good inner lining. The space between the studding is utilized for electric wiring and plumbing pipes. This type of construction has the advantage of better insulation than the traditional brick construction and is quicker to build.

Weather boarding, or tile hanging are suitable materials to use as cladding on timber-framed structures, but care must be taken in the detailing particularly at openings. Details are shown in diagram 57.

76 · BUILDING TECHNIQUES

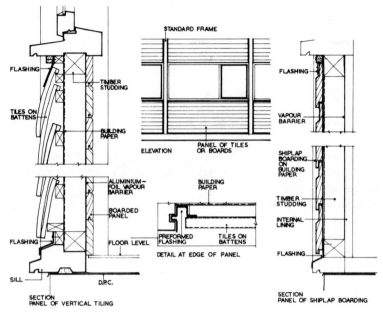

Diagram 57   Weather-boarding and tile-hanging

CHAPTER FIVE

Floors

Floors have to be strong enough to carry the usual furniture, people, goods, etc., expected in each type of building. For example, floors in houses are usually designed to carry safely 1·5 kN on each square metre of floor area, and warehouse floors should be able to carry 6·0 kN or more per square metre.

Timber floors are generally comparatively inexpensive and simple to construct. Concrete floors are more expensive and heavier, but are stronger and have better fire-resisting properties.

Timber
Timber for building is cut from trunks of trees (logs). A cut across a tree shows the way the fibres, which run up the tree, form a system of concentric annual rings, each ring having a soft part, spring growth, surrounded by a harder part, summer growth. In quick-growing trees the soft part of the annual ring can be quite wide, say 8 mm, and such timbers are usually softwoods (pines). The slower-growing trees have narrower annual rings and are usually hardwoods.

The tree grows by adding one complete ring each year to the outside. The outer part of the tree, known as 'sapwood', is therefore new and soft and the inner part, known as 'heartwood', is compact and harder.

Softwoods for building are available in long lengths and in a large range of widths and thicknesses (scantlings). They are easy to work and are cheaper than hardwoods, so are used for most carpentry, floors, roofs, partitions, etc., sometimes called 'carcassing'. Softwoods are also used for most joinery, doors, windows, cupboards, and some fittings, although most of the best joinery is done in hardwoods because a better finish is obtained in the closer grain of a hardwood and the figure, or pattern of the grain, is decorative. Hardwoods are generally more durable than softwoods.

Timber when cut contains much sap, mostly moisture and salts, which are drawn up in the sapwood. This dries out in time and the cells of the wood shrink as the water goes. It is essential that this shrinkage should take place before the wood is made into anything, and so the timber should be cut up and seasoned traditionally by leaving it for some years stacked in a manner which allows air to pass between the planks, protected from rain and sun, or by artificial seasoning in a kiln. It is now possible to measure the moisture content in timber with an instrument, so that it can be brought to the degree of dryness suitable for the chosen position in the building. Timber should not be too dry: in an unheated building it will absorb moisture from the air and may swell and split. The moisture content must be matched to the use.

If timber is not kept reasonably dry, it will rot. In the building it should be kept away from materials that are likely to be damp at any time, or the timber must be suitably protected by a preservative. 'Dry rot' is a fungus which grows principally on damp softwood. The spores or seeds are as small as dust particles and can lie dormant for years. Once dry rot grows it is very difficult to eradicate, as it sends out fine tendrils which pass through cracks in plaster and brickwork and so spreads to all the timber of a building. See also page 194.

The cutting up or conversion of a log is important, as the surfaces of the cut slabs will shrink more than the centres and they may become warped. The maximum shrinkage is in the direction of the annual rings; across the rings it is less, and along the grain it is very small indeed. This warping is shown in diagram 58, so that slab sawing, the cheapest method, produces a few good wide planks and several narrower less satisfactory ones. Rift sawing is better but more wasteful and expensive.

The most common defect in timber is the knot, which is a section of a branch. Most cheap softwoods have many knots and they have to be accepted, but timber with large loose knots should be avoided. A knot makes a break in the continuity of the fibres, so that on the underside of a beam it is a weakness similar to the notch shown in diagram 59. On the compression or upper side of the beam, or in its centre, it does not matter so much.

Cracks in the length of a timber, known as shakes, are also a serious defect often due to the tree being too old or to excessive drying.

Timber should, of course, always be cut up so that the grain is parallel with the edge of the plank. If the grain runs across at an angle, as it may at the end of a long plank, that part should be cut off and not used for structural purposes.

Softwoods used for carpentry and joinery are mainly:
    Redwood, imported Red or Yellow Deal and Norway Fir and home-grown Scots Pine and Fir.

FLOORS · 79

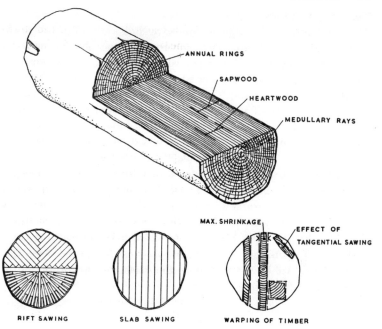

Diagram 58   Structure of timber and conversion of a log

Whitewood, including imported White Deal, Spruce, and Sitka or Western White Spruce from Canada and Norway — used for general work.

Douglas Fir, from Canada and the USA, including British Columbian Pine, Oregon Pine — hard and strong, with marked grain — used for general building construction.

Pitch Pine, from the USA — dark colour — used for durable joinery and structural work.

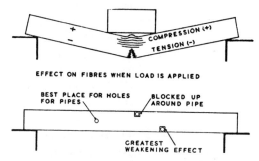

Diagram 59   Loading of a timber beam

Timber upper floors

The construction of an upper floor is really a problem of bridging or spanning the room below. The simplest way is to play beams across, bearing on walls or lintels and spaced a little apart. Each beam (joist) will carry an evenly distributed load which it takes from the floorboards bearing on it.

Diagram 60 shows a joist with a load indicated diagrammatically. With a small load (A) there will be a small deflection, not enough to notice, but some. If the load is increased (B), the deflection will increase and may become noticeable. If a much deeper joist (C) is used, it will not bend so much. The depth of joist necessary to prevent a noticeable deflection is therefore related to the load. If the same size of joist is used for a greater span (D) with the same load, there will be greater deflection again, but this will be less if a still deeper joist is used. So we see that the depth of the joist is related to span as well as load.

The load on one joist depends not only on its length and the type of use the floor will be put to, but on the spacing of the joists. But this spacing depends mostly on the floorboards, which are like shallow but wide beams spanning from one joist to another. In practice it is most

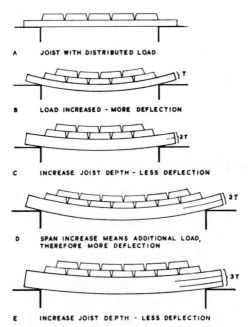

Diagram 60   Effect of load and span on the depth of a timber joist

convenient to use floorboards 19 or 21 mm thick and space the joists as wide apart as possible without getting deflection or 'springiness' in the boards. The usual spacing is usually about 400 mm centre to centre of joists for 19 mm boards.

Stress grading
It is necessary, if timber is to be used in an economical way, that its precise performance under load is known. This is not an easy matter because of the wide variety of species available and because of the defects to which the material is liable. Traditionally, the rule-of-thumb methods to determine the sizes of floor joists, etc., allowed an overgenerous margin of safety so that, in effect, a considerable amount of timber was wasted. Timber is therefore now graded in accordance with B.S. 4978 and following this the Building Regulations give figures for the size and span of joists suitable for floors.

The table below, compiled from the Building Regulations, gives suitable scantling sizes for floor joists using GS or MGS grade timber; this means visually graded or machine graded. Using these figures, the dead load on the joists must not exceed 25 kg/m^2. This allows the use of softwood or hardwood tongue and grooved boards, say up to 28 mm finished thickness. The floor will then be strong enough to support a superimposed load of furniture and people equivalent to 1·5 kN/m^2.

A width of 50 mm is shown in the table. Joists 63 mm or 75 mm wide are somewhat stronger than those 50 mm wide, but it is more economical to use a joist a little deeper and retain the 50 mm width. Much less timber is used thereby. A little more on the depth increases the strength considerably.

Sizes for Floor Joists

Size of joist (mm)		Spacing of joist − centre to centre		
width	depth	400 mm	450 mm	600 mm
		(maximum span of joist (m))		
50	75	1·35	1·22	0·93
50	100	2·25	2·03	1·58
50	125	2·84	2·75	2·33
50	150	3·40	3·26	2·84
50	175	3·95	4·78	3·30
50	200	4·51	4·31	3·76
50	225	5·06	4·83	4·22

82 · BUILDING TECHNIQUES

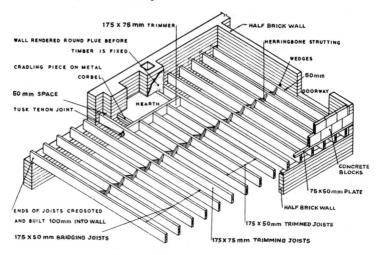

Diagram 61   Timber upper floor

Diagram 61 shows an upper floor. The room is 3·9 × 4·5 m. It is more economical to place the joists across the smaller dimension, as we can thus use 175 × 50 mm joists at 400 mm centres. They will be able to bear on a single brick (225 mm) wall on one side and on a half brick (103 mm) wall on the other; also the floorboards will then run along the length of the room, which is more economical and more pleasant visually. The layout of the joists has to be arranged to suit the fireplace, as no timber should go into the chimney breast; the carpenter has to form what is called a trimming. Trimming joists are placed 50 mm away from each side of the chimney breast and joists are positioned 50 mm away from the end walls. Between these joists others are placed to give approximately equal spacing and about 400 mm or less centre to centre.

The joists should have a good bearing at their ends; 100 mm is sufficient. At one end they rest on a 100 × 50 mm plate on top of the half brick wall; this plate is convenient for fixing. At the other end they go into the wall 100 mm and rest on the brickwork. This is satisfactory if the ends of the joists are creosoted and a little space is left at each side of each joist to allow air to enter and so ensure that it keeps dry. Wood plates used to be built into walls for the joists to rest upon, but they weaken the wall and are liable to rot. If the brickwork is laid accurately, the joists should all rest level; alternatively a metal plate 50 × 6 mm can be built into the wall. Joist hangers in the form of metal brackets projecting from the wall can be used if it does not interfere with the ceiling below.

To support the ends of the three joists between the trimming joists, a 'trimmer' joist has to be framed into the two trimming joists. A tusk tenon joint was traditionally used; a long tenon is formed on the trimmer and passes through the centre of the trimming joist and is secured on the other side by a wedge-shaped pin; this is a very strong joint, but now is largely superseded by a housed joint.

Each joist that the trimmer supports, known as a 'trimmed' joist, is fixed to the trimmer by a 'half-housed' joint, as in diagram 62, or alternatively may be notched over a fillet as shown. The latter if well nailed is a strong joint.

The hearth would be a cantilever hearth as described in chapter 3, diagram 44, but a member is needed to support the ends of the floorboards which come against the side of the hearth. This 'cradling piece' is usually a piece of floor joist notched or half-housed to the trimmer and carried at the fireplace end on a metal hanger built into the wall.

The trimmer may take a considerable load from the ends of the trimmed joists. So as to give it extra strength and allow for cutting away when forming the joints, it is usually made 63 mm wide, or

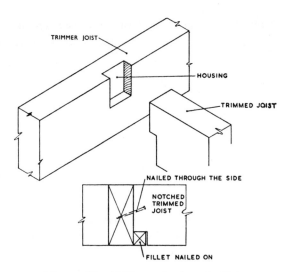

Diagram 62   Half-housed and notched joints

84 · BUILDING TECHNIQUES

75 mm if it supports six or more trimmed joists. The trimming joists certainly carry extra load and should be 75 mm wide.

The floor will be stiffened and any extra load on one joist will be spread sideways on to other joists if strutting is used. Herringbone strutting is shown in diagram 61. It is usual to place rows of this strutting so as to divide the floor into sections not more than 2 m wide. It is essential that wedges be driven between the end joists and the wall in line with the strutting as shown. Sometimes 'solid strutting' with pieces of floor joist is used; this is very good but not so easy to do well.

Openings of any shape for staircases, wells, etc., are trimmed on the same principle.

When a room is very large, say 8 m wide, it may not be possible to use joists spanning from wall to wall. The traditional solution here is to divide the area into smaller bays by using a structural steel beam to support the joists. These larger beams need a good area of bearing on the walls, as they collect a greater load from a larger area of floor. It is simplest to arrange them below the wood joists, with the latter on top. The joists can rest on a wood plate 100 x 50 mm bolted on top of the beam, but the latter will then usually project down too far for convenience and a more usual arrangement is shown in diagram 63. Here plates shaped to fit the beam are bolted to the web and on these the joists rest with their ends notched as shown to fit round the top flange.

To allow the ceiling plaster to be carried down the sides and across the soffit (underside) of the beam, timber cradling has to be formed as shown in the diagram; the metal lathing or plasterboard can then be fixed as on the ceiling itself.

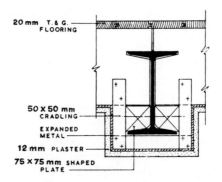

Diagram 63   Steel beam supporting timber floor

Reinforced concrete upper floors

Fire-resisting upper floors can be formed using concrete beams. A floor of hollow beams is shown in diagram 64. Each beam has its own reinforcement and is cast at the factory and hoisted up into place. Cross-reinforcement is laid over the beams and the whole covered with a layer of fine concrete. This is a good floor, but a proper bearing at each end is essential, the ends of the beams are easily damaged in transit and this will reduce the bearing area.

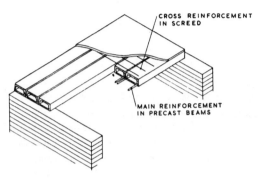

Diagram 64   Precast concrete floor beams

Formwork

The concrete floor illustrated in diagram 65 is shown cast *in situ*. That is to say, wooden platforms are erected at the level of the underside of the floor and propped in position by temporary supports. Where beams and columns occur, wooden moulds are made to the outline of the structural member. The steel reinforcement is placed on the platforms or in the moulds and correctly spaced out. This is done by wiring the reinforcement together and by using plastic formers or small concrete blocks as packing to ensure that the correct concrete cover is obtained. This temporary timbering is called 'shuttering' or 'formwork'. The technique is quite complicated, and is carried out by carpenters who usually specialize in this type of work. The construction must be strong since it must not only withstand the weight of the wet concrete, but also the impact loading as the concrete is poured into position. The joints must be tight and well made but at the same time the formwork must be capable of being easily removed when the concrete is set.

Since it is most economical to design formwork that can be used repeatedly, the timber is of heavy quality, and very often steel formwork is utilized and specially made where the construction unit can

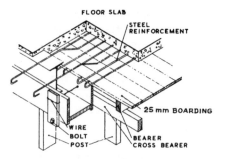

Diagram 65   *In situ* reinforced-concrete floor

be standardized. The formwork is left in position until the concrete has set, then it is removed or 'struck'.

The period of time that must elapse between pouring the concrete and striking the formwork will vary from one day for shuttering supporting the sides of beams and columns, to twenty-one days for props supporting main floor beams. The surface finish of the concrete depends upon the quality and texture of the material chosen to make the formwork. A variety of sheet materials are available for lining the soffit, the most common being 'extra-hard' oil-tempered hardboard, plywood, plastic-faced plywood and patterned rubber sheeting. A technique known as 'board finish' or 'board matted' concrete has been developed utilizing the texture of the natural timber used in the formwork. This is suitable where the concrete is exposed externally or for internal concrete walls and columns. The boards are carefully chosen and specially sawn so that the pronounced figure of the grain will be reproduced in the finished concrete. In order to prevent the concrete adhering to the formwork as it dries out, a non-staining mould oil or cream is applied in a thin film to the inside surfaces of the materials before the concrete is poured.

As an alternative to the finish direct from the formwork, attractive effects can be obtained by the removal of the vein of hard cement which forms on the surface of cast *in situ* concrete. This can be done by 'bush-hammering', tooling, chiselling or grading and polishing a surface. It is usual with these techniques to paint the inner surface of the formwork with a retarder to delay the setting of the surface of the concrete and make it easier to produce the surface texture. The area of cross section of reinforcement necessary to take all tensile stresses in the concrete has to be calculated and the number and size of rods chosen to give this area. Additional reinforcement is required to take

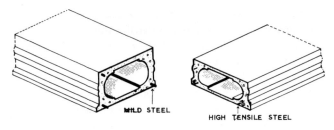

Diagram 66　Comparison of precast and prestressed concrete beams

other stresses set up, and in all but the simplest cases a Structural Engineer should design the whole floor slab. But as a guide we can say that a floor having 12 mm rods at 150 mm centres placed 15 mm from the bottom of a 150 mm concrete slab will span 4 m and be suitable for the floor of an office.

There are other types of floors, some using steel lattice beams as shown in diagram 67, which are now commonly used for medium-span floors from 4 to 8 m, as in schools.

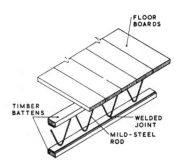

Diagram 67　Metal lattice floor joist

Pre-stressed concrete beams

Beams for flooring (and, of course, for roofing) can now be obtained in 'pre-stressed' concrete. This means that the concrete is stressed before it is used in its position in the building. High-tensile small-diameter cold drawn steel is used as reinforcement. The wire will not be more than 6 mm diameter compared with 12 to 20 mm diameter rods of mild steel used in ordinary beams for the same loading. The principle of pre-stressing is to stress the concrete in manufacture so that when it is under load and in position the tension artificially induced in the beam balances the tension produced by the superimposed load. Concrete may be either 'pre-tensioned' or 'post-tensioned'. Pre-tensioning is very suitable for factory production and so this is the technique used in the manufacture of pre-stressed floor and roof beams. The reinforcing wires are first stretched against fixed abutments and then concrete is poured and allowed to harden before the wires are cut. Pre-stressed beams usually have a slight camber or 'arched' profile along their length due to the tension set up by the wires. Where post-tensioning is used the wires are stretched and anchored after the concrete has set. This is done by passing the wires through holes left in the concrete during manufacture. Post-tensioning is more often used *in situ* in structural engineering projects.

Pre-stressed beams are more expensive than ordinary reinforced beams but they are smaller in section and therefore lighter in weight per unit of load to be carried. This means a considerable saving in the cost of the structure to carry the beams in a multistorey building. Diagram 66 compares the cross-sectional areas of a typical ordinary pre-cast beam with a pre-stressed beam. Both beams would carry the same load over a given 'span'. Where pre-stressed work has to be demolished, extra care must be taken due to the inherant tension in the structural members.

Ground floors in timber

Where boarded ground floors are required a similar construction to the upper floor can be adopted, except that intermediate support can be got from the surface concrete. Half brick walls can be built up as shown in diagrams 68 and 69 wherever convenient. We can therefore choose a manageable size of joist, 100 x 50 mm or 125 x 50 mm, and space the sleeper walls accordingly.

Diagram 69 is a typical plan of a ground-floor room in a small house. From the table of joist sizes (page 81), it can be seen that a 75 x 50 mm joist would need two sleeper walls across the middle of the room, but 100 x 50 mm joists will span more than halfway needing only one sleeper wall at the centre. In some districts it is not usual to have separate sleeper walls at the sides, the joists quite often being built into the main or partition wall, but it is good practice to build separate sleeper walls. They do not take much weight and so, to ensure ventilation and economize on bricks, they are built in the manner shown in diagram 68 and are called honeycomb sleeper walls.

FLOORS · 89

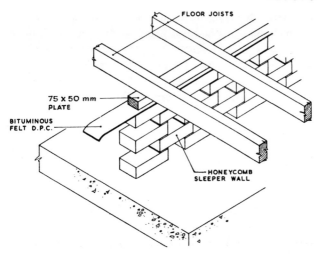

Diagram 68   Honeycomb sleeper wall

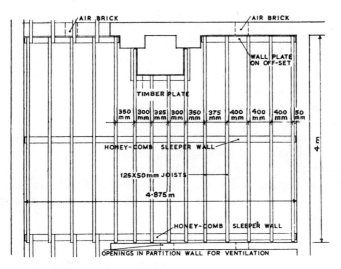

Diagram 69   Timber ground-floor construction

Site concrete
The ground under a suspended timber floor should be covered by a 100 mm layer of good-quality site concrete. The Building Regulations require that the top of the concrete must not be below the outside ground level. This is to prevent water lying over the surface of the

concrete where site conditions are bad. In the past, where ground floors in timber have been badly constructed so that the site concrete was placed below outside ground level, the dampness caused by the standing water has been a major cause of decay in the timber.

The site concrete must be laid on a bed of hardcore which must be clean and free from sulphate impurities which could pass into solution and damage the concrete.

Underfloor ventilation

The Building Regulations allow the space between the top of the site concrete and the underside of the wall plate to be reduced to a minimum of 75 mm, provided that adequate ventilation can still be obtained. It is probably better, however, to build a two-course honeycombed sleeper wall as shown in diagram 70. This means that the d.p.c. can be laid at the same level throughout which simplifies the damp-proofing.

The space between the floor and the timber is ventilated by means of small ducts through the cavity walling. The ducts are formed by roofing slates which bridge the cavity. The slate is placed around the airbrick which fits into the brick courses on the outside face of the wall. The diagram also shows the details at a partition wall, where it will be seen that holes are left to allow a through draught which ensures satisfactory ventilation.

Diagram 70 also shows a suspended timber floor built next to a solid floor. This construction often occurs and special precautions must be taken to make sure that cross-ventilation can take place. This is done by laying drainpipes to act as airducts to connect the space underneath the timber floor to the outside air. The pipes are terminated by a standard bend to bring the outlet as far above the ground level as possible and then an airbrick can be used in the outside leaf of the cavity wall to complete duct outlet.

Floorboarding

In domestic work the tongued and grooved softwood boarding is usually laid as an integral part of the suspended timber floor construction and so the question of a 'floor finish' does not arise. The householder will probably cover the boarding with carpet or linoleum which should be laid loose so that the floor boarding is able to 'breathe'. For cheaper work, therefore, where appearance does not matter, the boarding in planks 100 mm or 125 mm wide would be surface nailed. The pattern of the nailing will be seen along the lines of

FLOORS · 91

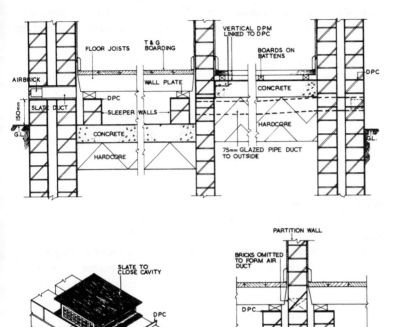

Diagram 70 Details of underfloor ventilation to suspended timber ground floor

the supporting joists, but the nails should, of course, be punched below the surface as shown in diagram 71. If, however, the boards are to be stained and polished so that they remain exposed or if hardwood strips 50 or 75 mm wide are used, then the boarding should be secret nailed. Oak, Beech, Sycamore, Mahogany and Teak are all excellent for use as strip flooring, but there are many more species from which to choose. In this method the nails are driven at a slight angle into the tongues of the board or strip as shown in diagram 71.

Note should be taken of the fact that the tongues contribute significantly to the strength of the flooring. The joists would have to be closer together if plain-edge boards were used. Sometimes a hardwood border may be laid in a room leaving the centre portion in softwood to

92 · BUILDING TECHNIQUES

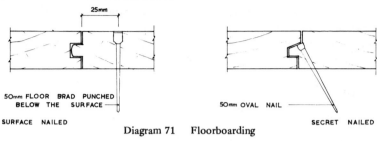

Diagram 71   Floorboarding

be covered by carpet. If this is done, the hardwood can be slightly thicker so that the carpet locates easily and does not kick up at the edges.

Solid ground floors

Where the site is level, a simple form of floor construction laid direct to the ground may be used. This will consist of a bed of hardcore upon which a concrete slab is laid to receive a floor.

The hardcore should 100 mm or 150 mm thick, consisting of clean broken bricks, graded, blinded with fine material and well rolled, or clean broken natural stone well rolled and consolidated. Builders' rubbish and particularly old plaster should not be used for filling since it will probably contain deleterious salts. The hardcore is to provide a good, well-drained foundation for the concrete.

The concrete should be a minimum of 100 mm thick and be of good quality. A mix of 1 part cement, 3 parts sand and 6 parts coarse aggregate will be adequate. The concrete is usually specified as 'spade finish', that is to say smoothed over with the back of a spade to receive the screed and then the floor finish.

The screed is laid to form a true and level surface to receive the finish, the best mix for the screed being 1 part cement to 3 parts sand mixed and laid as dry as possible. The screed must be minimum 20 mm thick when laid direct on to the concrete, and a minimum of 50 mm thick when laid over a damp-proof membrane. The thinner screed is bonded to the concrete; but the thicker screed, because of the membrane barrier, must rely on its own weight and thickness to remain intact. The finish for a solid flooring will probably be one of the following: PVC vinyl (asbestos) floor tiles; clay floor tiles; quarries; cork tiles; Terrazzo in tile form or *in situ*; woodblock flooring (w.b.f.); parquet; rubber sheet or tile flooring; linoleum in sheet or tile form; timber strip flooring on timber fillets.

Where the floor finish is adversely affected by moisture, or where timber is used, it is necessary to provide a damp-proof membrane. This may be a continuous sandwich of hot bitumen 4 mm thick, or three coats of bitumen-rubber solution sandwiched between the concrete and the screed to receive the floor finish.

Woodblocks should not be less than 16 mm thick and must be dipped in bitumen so that a continuous waterproof layer is formed below the block. Alternatively, they can be laid in special adhesive on a 12 mm layer of asphalt.

Where strip flooring is used, it is permissible to lay the timber fillets, which must be impregnated with preservative, on top of the d.p.m., and fix them by clips into the screed.

Wherever a d.p.m. is used it must be at least 150 mm above the outside ground level, and it must be continuous with the walling damp-proof course. It is now common practice for polythene sheet used as a damp-proof membrane, to be laid on top of the hardcore, underneath the site concrete. The polythene can then very easily be brought up the walls to form a vertical d.p.m. equal to the thickness of the concrete floor slab, and then be tucked into the brickwork just below the horizontal d.p.c. to form a continuous water barrier. The advantage of using polythene is that it can be obtained in wide rolls, usually 4 m, which reduces the number of lapped joints. A disadvantage is that it is fairly easily punctured by any of the innumerable sharp objects that find their way on to a building site, and for this reason it should not be too thin, certainly not less than 0·3 mm. Diagram 70 shows details of solid floor construction. Because of their nature, certain floor finishes such as clay floor tiles or quarries, do not need, in theory at any rate, a damp-proof membrane.

The relative position of screed and membrane is illustrated in diagrams 20, 21 and 23 in chapter 3.

Condensation

The provision of hardcore under a solid floor helps to prevent ground water rising through the concrete. It does not, however, prevent the concrete becoming damp by the absorption of water vapour; but provided that the floor covering is permeable, as with clay or concrete tiles, the moisture can evaporate into the room. On the other hand, the presence of the water vapour may encourage the formation of condensation on the cold floor surface. This condensation would be prevented by the damp-proof membrane previously described and for this reason it is probably better to have a waterproof membrane in every case.

CHAPTER SIX
Roofs

Structure

The complexity of the roof structure is related to the span required. The larger the span, the more complex the structure will be. Diagram 72 sets out in very simple terms the wide range of structural forms and covering materials that are available.

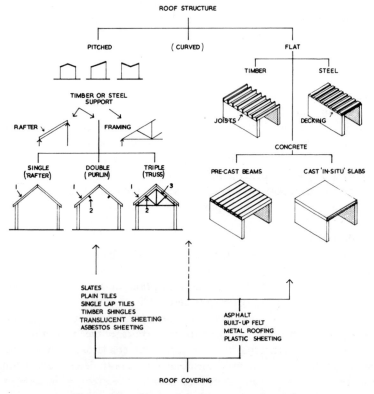

Diagram 72   Forms of roof structure and coverings

Insulation

It is sound economy to ensure that a roof structure is properly insulated and the Building Regulations stipulate that the thermal transmittance through the roof of a dwelling shall not be more than $U = 0.6$, which requires a very high standard of insulation. For industrial buildings the roof insulation required is related to the internal temperature of the building. The higher the temperature, the lower must be the heat loss. Note that the indiscriminate adding of insulation to a construction may increase the risk of condensation forming within the construction, due to the increased difference between the inside and outside temperatures of the building fabric. Therefore, care must be taken in the provision of correctly placed vapour barriers (see page 98) together with the consideration of adequate means of ventilation to the construction in particular to timber roofs.

Diagram 73 compares an uninsulated roof with an acceptable standard of insulation both for concrete and for timber construction for an internal temperature of $15°C$ in a factory building. The U value at this temperature must not be greater than 2.4.

Fire resistance

Note should be taken that the addition of insulation should not increase the risk of the spread of fire within a building. Materials which

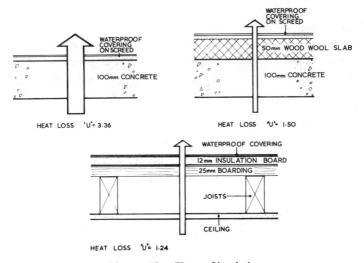

Diagram 73   Flat roof insulation

are commonly used in the manufacture of insulating quilts and boards by their nature are usually of a cellular or fibrous makeup utilizing the trapped still air as a barrier against heat transfer. Unfortunately this air provides the oxygen for combustion where the walls of the cells are themselves combustible. It follows, then, that the entire material must be rendered fire-resistant, or else the surface must be treated to resist the spread of flame across it. B.S. 476 describes certain tests to prove the reliability of surface treatments and gives three classes of resistance. Wherever possible, materials having a Class I rating to surface flame spread should be used.

Wind-loading on roofs
Effect of wind-loading on roofs is complicated and depends on the angle of pitch and the degree of exposure of the roof position. The wind effect to be guarded against in detailing is the amount of suction that may be expected. Between 20 degrees and 30 degrees of pitch the suction is equally balanced on both sides of the roof, and on steeper pitches the suction takes place mostly in the leeward side. For shallow pitched roofs up to 20 degrees, there is more suction on the windward side. With heavy traditional roofs, suction does not present a problem; but with shallow pitched roofing with lightweight covering, it is a wise precaution to make sure that the roof is anchored down. This should be done by securely spiking the joists to the wall plate, and then securing the wall plate by means of metal straps built into the brickwork.

Snow-loading
Roofing must be able to withstand the weight of snow, and also be able to carry the loading due to workmen and materials when maintenance is being carried out. An allowance of 75 kg/m^2 is usually made.

Flat roofs
Materials for coverings
The materials for flat-roof coverings fall into two classes. These are skin coverings such as asphalt or felt and bitumen, and sheets, such as lead, copper, zinc, etc.

The sheet methods were well developed by traditional craftsmen. The metals used do not corrode and are pliable in thin sheet form, so that joints can be made by folding the edges together in such a way that rain runs quickly off the fold before it can be carried between the folded surfaces, either by pressure or by capillary action. This means that it is more important to shed water quickly from a sheeted roof than from a skin-covered roof. For sheeted roofs, therefore, firrings to falls, described later, are necessary and mean additional timber and much complicated labour. Asphalt and bituminous felt roofs show savings in material and labour, although they are not regarded as making such efficient coverings where long life and complete reliability are concerned.

Asphalt
Asphalt is a mixture of a bitumen binder and a filler, rather like a bituminous mortar. A typical mixture would be bituminous asphalt with crushed asphalt rock. When heated the mixture softens and can be spread with a trowel. Two coats are usually laid, to a total thickness of 20 mm. It can be dressed easily round pipes and irregular shapes, etc.

Asphalt can be laid on boarding, but as some movement may take place in the timber the asphalt should be isolated by laying a sheathing felt under it.

'Built-up' felt
Bituminous felt is prepared in rolls. It is a fabric of hessian, jute or asbestos or glass fibres coated on each side with bitumen and some harder-wearing material for a surface. The roof is covered with layers of this felt, with hot bitumen between.

Built-up bituminous felt roof systems are particularly suitable for use on wood roofs, as there is a small degree of flexibility in the sheets themselves and in the bitumen layers between them; therefore they are less likely to crack.

It is important, however, that only good-quality felts are used and that the work is carried out by reputable firms. Felt is difficult to dress round pipes and irregular shapes, where it has to be cut and lapped in a complicated way.

Vented underlays
Another disadvantage of built-up felt roofing systems is that heat from the sun's rays, particularly in changeable weather, may vaporize trapped moisture which builds up pressure between the layers of felt causing the surface to blister and finally to crack. This can be overcome by using a base layer having graded granules on the underside which permits the built-up vapour pressure to be dispersed by means of specially designed flashing details as shown in outline by diagram 74.

98 · BUILDING TECHNIQUES

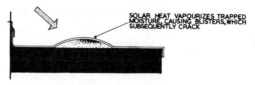

FELT BONDED DIRECT TO SUBSTRUCTURE

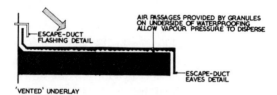

'VENTED' UNDERLAY

Diagram 74   Ventilated built-up felt roofing

Promenade roofing
Where flat roofs are to be used as promenades, it is necessary to surface them in a material that will withstand foot traffic. This can be done by laying asbestos cement tiles approximately 300 mm square on top of the waterproofing. The tiles also will reflect a high degree of solar heat, thereby reducing the risk of structural movement. These tiles are shown in diagram 80 together with a detail of a balustrade to act as a safety barrier around the perimeter of the roof. The balustrade consists of 25 mm square hollow, mild-steel sections built into a concrete kerb. The handrail is screwed up from the underside through the steel core, and the horizontal rails are bolted to the steel uprights in a similar way to the detail of the stair balustrade in diagram 151. The asphalt is carried over the concrete kerb and is tucked into a groove behind the slate facing to protect the edge of the roof. This facing is cramped back into the concrete by small copper cramps.

Vapour barrier
Where conditions are warm and humid inside a building, there is a danger that the warm damp air will move through the roof structure and condense within the roof or wall thickness (interstitial condensation). This moisture may cause the structure to deteriorate and in severe cases the moisture will drip back into the building. The temperature at which condensation takes place is called the dew point.
  In order to restrict this vapour movement, it is necessary to provide a vapour barrier on the warmer side of the insulation. The vapour barrier should be placed to prevent the moisture vapour getting to a

point within the construction where the temperature drop would otherwise cause the vapour to form condensation. Diagram 75 illustrates the principle. The barrier may be two coats of bituminous paint or a layer of bituminous building paper well lapped and sealed at the joints. If a sheet material is chosen, it is important that the membrane is continuous and that all joints are sealed. If humid conditions make it necessary, a vapour barrier would also be incorporated in the construction immediately underneath the insulation shown in diagram 82.

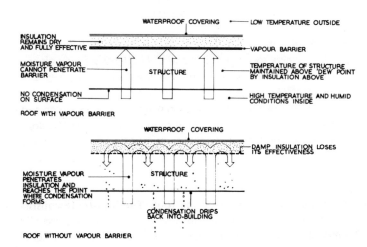

Diagram 75   Vapour barrier

Metal sheet roofing

Metal sheets for roofing may be of lead, copper or zinc. Lead has been most widely used in Britain in the past, but if the raw-material price of lead and copper are comparable, there is little to choose between these materials.

Sheet lead is rolled out to an even thickness. The most usual thickness is 2·24 mm, Code no. 5, which weighs approximately 25 kg/m². Lead is malleable, i.e. easily beaten out to a shape, and ductile, i.e. it can be drawn out without appreciably reducing the thickness; a flat sheet of lead can be beaten out to cover a hemisphere. Allowance must be made for thermal expansion, which is appreciable.

Copper sheets are usually 0·9 x 1·8 m long. The thickness used for good-class work is 0·6 mm (23 Standard Wire Gauge) which shows a considerable saving of weight over lead. For cheaper work, thinner (0·45 mm 26 swg) sheets can be used. It can be dressed to form angles, but gutters, ends of rolls, etc., have to be made by folding as the sheets are so thin. This makes the dressing to irregular shapes rather complicated and much skill and experience are needed. Copper expands only a little for normal rises in temperature and it will return to its original shape. It weathers to a pleasant green colour within a few years.

Zinc for roofing is used mostly in 14 Equivalent Zinc Gauge, which is 0·8 mm — which is equal to 21 s.w.g. and weighs 5·76 kg/m^2. It is cheaper than lead or copper but not regarded as so suitable for town atmospheres. The methods of jointing and working are very similar to copper.

The principle of constructing a sheet-metal roof is based on the fixing of the sheets to allow for expansion, either by fixing along two adjacent edges as with lead, or by holding down with clips as with copper or zinc. The joints of sheets across which water flows must be joined on a step down or drip, shown in diagram 87; joints across which water does not flow can be made waterproof by turning up and wrapping over a wooden core or roll, as in diagram 86.

Compound sheet roofing

A compound roof material of thin (42 g) copper sheet with a backing of bituminous felt is available as a less expensive alternative to traditional copper. The bituminous copper roofing is supplied in rolls 600 mm wide. The sheet must have an underlayer of built-up felt. The finished roofing has something of the appearance of traditional copper. Side joints running with the slope of the roof can be standing seams, welted seams, or batten rolls, and all cross-joints are welted.

Thermoplastic sheet roofing

An addition to the range of materials available for roof covering is a thermoplastic sheet of asbestos and bitumen laminated in waterproof sheet form, standard size 2·4 x 0·9 m. It is less expensive than the roofing metals and can be used in similar situations. It must always be laid on a solid backing and be fully supported, and have a fall at least 20 mm per metre. Thermal expansion is less than that of zinc or lead but slightly greater than that of copper. Fixing technique is similar to that used in sheet-metal roofing and the roll-cap method is the most satisfactory. This material has the additional advantage that when the sheets are heated with a blowlamp they become pliable and welded joints can be made. This is done by splitting the laminations on the edge of the heated sheet for say 25 mm in depth, inserting the next sheet, coated with fibrous welding plastic, and sealing down again with a hot iron.

Typical joints used in sheet roofing are illustrated in diagram 76. It should be noted that joints parallel to the eaves will be folded down as flat as possible to allow the water to run down the roof easily, while joints which follow the slope from ridge to eaves have an upstand to guide the water more quickly down to the gutter.

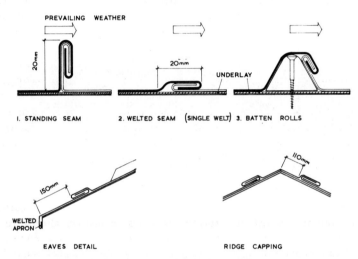

Diagram 76   Joints in sheet roofing

Falls to flat roofs
A fall of 20 mm in 1 metre is recommended for sheet-metal coverings, and a similar fall is also the minimum for asphalt and felt roofing, although a greater fall is desirable wherever possible. For bituminous felt roof coverings, some experts have advocated doing without a fall, which saves the cost of additional material, and argue that the damp surface preserves the asphalt or felt from excessive heat in summer. This argument, however, is a false one since unless the structure is designed to retain at least say 25 mm of water the surface will not remain damp, but will dry out in patches. This is because most sheet materials in use today as roof decking do not in practice remain level and true after laying, and even the slightest depression in the roof deck will cause shallow pools of water to form on the finished roof surfaces. This defect is known as 'ponding' and is a notorious cause of leaking roofs.

The leak is caused by the excessive thermal movement which takes place around the edges of the damp area causing the asphalt or built-up felt to crack.

The arrangement of the fall depends on the plan, possible positions for rainwater pipes, positions for gutters and the type of roof construction. It is, however, possible to have a flat roof without a fall if the surface is protected from excessive thermal gain from the rays of the sun by at least a 50 mm thickness of loose gravel. Drainage must be by a 'weir' system to prevent the roof flooding but at the same time keep the chippings damp for as long as possible, and of course account must be taken of the additional weight of wet gravel.

Inverted roof

Sheet insulation materials have been developed which are waterproof and so can be placed on top of the bitumen or asphalt. These materials also protect the waterproofing from excessive temperature variation and so reduce the risk of cracking. The insulation must, of course, be loaded with a good thickness of gravel (50 mm) to keep it in place. The technique of placing the insulation *above* the waterproofing has become known as the 'inverted' or 'upside-down' roof. Trouble can in some cases be caused by mould growth around the insulation.

Concrete flat roofs

On concrete roofs, the fall is usually made by laying a cement and sand screed over the whole of the structional roof slab. This screed will be 20 mm minimum thickness at the lowest point; if it is thinner than this, it will not adhere to the slab.

On a rectangular roof the gutters are best placed along the two long sides. Then the fall is short and screed thickness is saved. But gutters can be on two adjacent sides, as in diagram 77. Parapets protect the other two sides. Here the fall is in two planes which intersect on the diagonal line, though in fact the change of plane is so small as to be imperceptible. Note that the rainwater pipes are spaced evenly round the building. In diagram 78 the gutter is on one side only, and consequently there is one fall only across the roof. The rain must be prevented from blowing over the exposed edge, so if there is no parapet a curb must be formed as in diagram 80. Note also the use of an *in situ* concrete eaves detail in conjunction with a hollow, pre-cast reinforced concrete beam. When parapets are on all sides as in diagram 79, gutters can be formed in the flat roof or the falls can be arranged to one or more points. In this plan there is only one outlet provided and a 'valley' is formed between the two planes of screed to direct the water to the outlet through the parapet.

ROOFS · 103

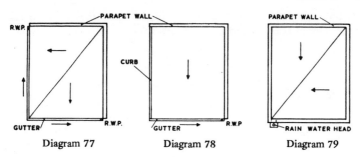

Diagram 77   Diagram 78   Diagram 79

Falls and gutter positions for flat roofs

An outlet through a parapet well is shown in diagram 81. The asphalt can be dressed through and to the sides of the opening. A sheet of lead can be fixed to hang down into the rainwater head, and the asphalt is dressed over this. Note the insulation, which may be of fibreboard or wood wool or other suitable material.

A similar lead apron is used at the edge of a roof where a gutter is fixed, as in diagram 82. This also allows the gutter to be replaced without disturbing the asphalt.

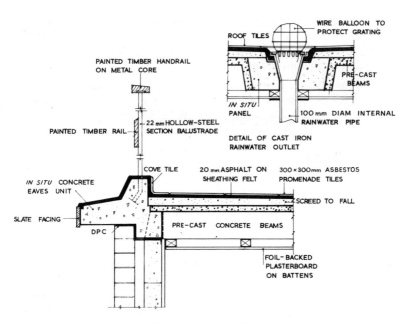

Diagram 80   Eaves of concrete promenade roof – asphalt covering

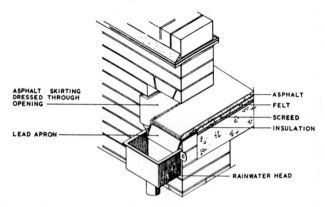

Diagram 81   Rainwater outlet through parapet wall — asphalt covering

Where there is no gutter an edge of even thickness can be arranged by using a curb and sloping the surface as shown in diagram 80. Only a little rain will drip off the part sloping outwards. Also illustrated is a roof outlet fitted to an internal rainwater pipe. This type of outlet can be obtained either in cast-iron as shown or PVC.

An asphalt skirting against a parapet wall is shown in diagram 37. It is important that the chase into which the top of the skirting is pressed is made deep enough and square in section. If skirtings are high, or where a whole wall has to be covered with asphalt, reinforcement, usually expanded metal lathing, should be used, otherwise the asphalt will soften in the sun and sag downwards.

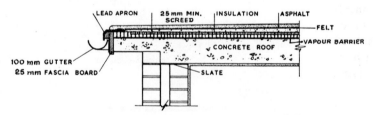

Diagram 82   Eaves of concrete flat roof

Timber flat roofs
The principle of providing the fall is the same for wood roofs, but only the more simple 'one-way' arrangement of falls to opposite sides are usually possible, as the fall has to be made by fixing tapering pieces along or across the top of the joists. These are called firring pieces and

ROOFS · 105

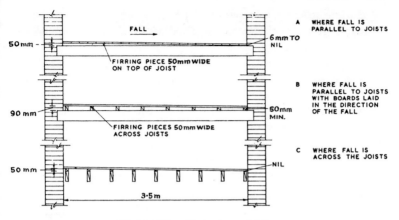

Diagram 83   Firring to roof falls

are shown in diagram 83. The simplest arrangement is shown at A. Here each joist has a piece fixed on top of it tapering from approximately 50 mm to 6 mm. The boards above this will run across the slope, which is not good; if the boards warp, ridges are likely to appear in the asphalt and obstruct the flow. A better arrangement is shown at B, where cross-firrings are used. As these have to support boarding and span from joist to joist the smallest member must be 50 mm minimum depth. This makes the one at the top of the fall 85 mm deep in this case, using much more timber. Where the fall is across the joists, as at C, each firring piece is a different depth. This is easily done in this case by using 50 × 50 mm timbers sawn so that the offcut can be used for one of the thinner firrings.

An eaves detail with gutter is shown in diagram 84. To reduce the depth of fascia, the flat-roof joists are notched. So that the boarding can be parallel to the slope, the 50 × 50 mm firrings are placed across the joists. The three layers of bituminous felt are dressed to stop at the

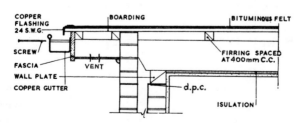

Diagram 84   Flat timber roof with eaves gutter

edge of the boarding, a copper flashing being fixed to the boarding first. A copper gutter as shown can be fixed with long screws through the tubular spacers in the gutter.

An alternative form of timber flat roof with a wood-wool slab decking is shown in diagram 138. Firring pieces are used to give the necessary falls to the roof but in this example they are placed along the top of the joists. The wood-wool slab is screeded in 1 : 5 cement screed to provide a smooth surface to receive the felt. Another effect of the screed, which keys into the surface and into the joints between the slabs, is to help to make the construction more rigid and so reduce the risk of subsequent failure of the built-up felt due to movement of the substructure.

The built-up felt is in three layers, the upper layer being asbestos based, to B.S. 747, in order to provide fire protection. The sheets are bonded in hot bitumen and are laid parallel to the gutter and with lapped joints breaking bond. The top or cap layer may be mineral-surfaced with small self-coloured granules, or white spar chippings may be scattered over the surface and set in bitumen to reduce solar heat absorption and to prevent the ultraviolet rays from the sun attacking and breaking down the covering. The diagram indicates the use of chippings but care must be taken that these are rounded so that they will not pierce the felt when walked on during maintenance work.

Lead covered flat roofs
Diagram 85 shows a lead flat. The position of the outlet has dictated the arrangement for the fall. With joists spanning the short way, a fall can best be obtained by placing firrings of different depths on the tops of the joists.

Strips of lead 2·4 m long can be cut conveniently from a roll, so that

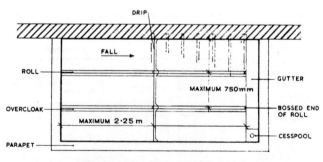

Diagram 85   Plan of lead flat roof

drips are necessary at approximately 2·25 m centres. For ease of handling sheets are usually cut about 0·9 m wide also strips wider than this would cause excessive movement at the roll when they expand. Each sheet is fixed on one side by nailing with copper nails to a wood roll (diagram 86) and the sheets nearer the gutter are similarly fixed at their higher end, where they are dressed under the next row of sheets. Where sheets abut against the parapet wall, they are turned up as shown in diagram 87. The upper edge of this turn-up is protected by a separate cover flashing, a strip of lead turned into a joint of the brickwork and wedged and pointed. This has to be stiffened at intervals by lead tacks, strips of lead at the back wedged into a joint in the brickwork and gripping at the bottom the edge of the flashing. The gutter is formed by lining with lead turned up and dressed under the roof sheets and against the wall. The details are similar to those described above. Long gutters have to have drips at intervals.

The outlet to the rainwater pipe is often formed to a cesspool, or deeper part of the gutter, which is separately lined with one piece of lead, the gutter draining into the cesspool with a drip.

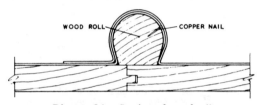

Diagram 86   Section of wood roll

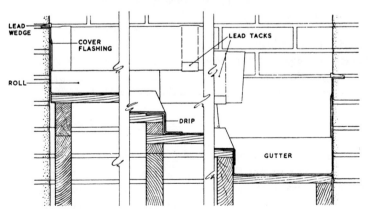

Diagram 87   Lead roof – sections through drips and gutter

Pitched roofs

When rain falls on a pitched roof the water will fan out and run over the surface at a given angle. This angle will depend upon the pitch of the roof and is sometimes referred to as the 'angle of creep'. The steeper the pitch, the narrower the angle will be and this is a guide to the minimum width of the covering unit that can be used. Thus it follows that the shallower the pitch, the wider the covering units will have to be. Diagram 88 shows the relationship of the pitch to the choice of roof-covering material.

Due to the method of hanging tiles and slates on battens, the effective pitch at which the tiles are laid is less than the pitch of the rafters, so that the rafter pitch should always be greater by about 5 degrees than the minimum pitch recommended for each unit.

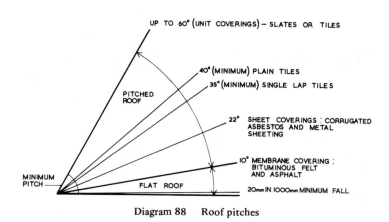

Diagram 88   Roof pitches

Materials for coverings

Diagram 89 compares the techniques of laying plain tiles, single-lap tiles, interlocking tiles, and slates. There are several widely used proprietary variations in profiles available for the single-lap interlocking tile.

Plain tiles

These are small slabs of burnt clay (or concrete) 265 mm long x 165 mm wide x 10 mm thick. They are hung and positioned on the battens by means of the projecting ribs and are fixed by 30 mm 'composition' or aluminium nails. Composition is an alloy of copper and zinc with a very small proportion of tin. It is not necessary to nail

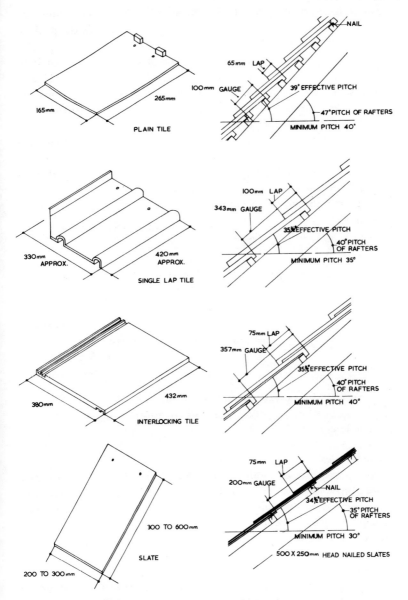

Diagram 89   Comparison of tiling and slating

each tile in each course unless the pitch is over 60 degrees and the exposure is severe. An adequate specification for 40-degree pitch with moderate exposure would be to fix by one nail per tile every fifth course. Plain tiles are laid with a 'double-lap' and have three thicknesses of the tile over the point of fixing. Tiles are laid 'half-bond' and butt-jointed.

Before setting out the tiling on the roof the distance from centre to centre of the battens must be worked out. This distance is known as the 'gauge'. The 'gauge' depends on the overhead lap at the head of the tile and this is determined by the degree of exposure on the site. A lap of 65 mm is suitable for tiles of average exposure and the gauge is worked out from this as shown in the following example:

First choose lap:

i.e. 65 mm for normal exposure:

Then

$$\frac{\text{length of tile} - \text{lap}}{2} = \text{gauge}.$$

Thus

$$\frac{265 - 65}{2} = 100 \text{ mm gauge}.$$

Length of the tile at eaves = gauge + lap
= 100 + 65 mm
= 165 mm eaves course,

i.e. special tile at eaves 165 mm long × 165 mm wide.

Verge course is always 'tile and a half' in width giving a special tile 265 mm long × 248 mm wide.

Single-lap tiles

Single-lap tiles overlap both at the sides and top so that they do not require to be laid double-lap. The double Roman type of tile shown in diagram 89 is an example of this. Most single-lap tiles are larger units than plain tiles, say about 420 mm in length. This means that battens can be at a much wider gauge, as is shown in the diagram.

Interlocking tiles

Interlocking tiles are also laid single-lap but are designed to have a fully interlocking joint at the sides, so that there is no change in profile at this joint.

Slates

Slates are split from rock which is quarried in Wales and the Lake District. A large range of sizes of slates are available, the following (given in millimetres — rationalized metric sizes) being the most used:

$$600 \times 300, \ 500 \times 250, \ 450 \times 250 \text{ and } 400 \times 200.$$

Because of the 'angle of creep' the slates are laid to steeper pitch as the width decreases, thus:

200 mm wide: minimum pitch 35 degrees
250 mm wide: minimum pitch 30 degrees.

In general although there are exceptions, Welsh slates are dark blue-grey and Lakeland slates are grey-green. Slates may be centre-nailed or head-nailed (holes 25 mm from the top of the slate). Centre-nailing is preferable for larger sizes to prevent lifting by the wind, and is the most used technique.

Slates, like plain tiles, are laid double-lap, with special slates at the eaves and slate-and-a-half at the verge. A 75 mm lap will be found to be adequate under normal conditions. Thus the gauge may be worked out as follows, for centre-nailed slates:

First choose lap:
i.e. 75 mm for normal exposure:
Then

$$\frac{\text{length of slate} - \text{lap}}{2} = \text{gauge}.$$

Thus for 500 x 250 mm slates:

$$\frac{500 - 75}{2} = 212.5 \text{ mm gauge.}$$
(say 212 mm for practical purposes)

If the slates were to be head-nailed, as shown in diagram 89, the fact that the nail holes are punched 25 mm from the top of the slates must be taken into account.

Thus
75 mm lap:
500 x 250 mm head-nailed slates:

$$\frac{\text{length of slate} - (\text{lap} + 25 \text{ mm})}{2} = \text{gauge}$$

112 · BUILDING TECHNIQUES

i.e.

$$\frac{500 - (75 + 25)}{2} = 200 \text{ mm gauge.}$$

The battens upon which the tiles and slates are fixed should not be less than 38 mm wide. They should be of sufficient thickness to prevent undue spring under the hammering, and with rafters at 400 mm centres, 19 mm will just be adequate but 25 mm will be preferable.

Before the battens are fixed, the spars are covered with reinforced slaters' felt; this is laid starting from the eaves with all joints lapped 150 mm. The felt should sag slightly between the rafters (in order to take away any water which may blow under the tiling or slating) and should be carried over the fascia board far enough to give a 'drip' into the gutter. A specially shaped section of the timber known as a 'tilting' fillet should be used in all cases to close the gap between the rafters and the tiles or slates at the eaves.

Systems of pitched roof construction

Some of the common terms used in roof construction are shown in diagram 90.

We have seen that the covering material decides the pitch or inclination of the roof members. These inclined members or rafters take the load from the battens and pass it to the plate which is fixed to the top of the walls or to other framing members in the roof. Each rafter acts therefore partly as a beam and partly as a strut. When the load

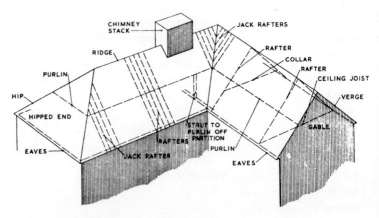

Diagram 90   Pitched roof construction − common terms

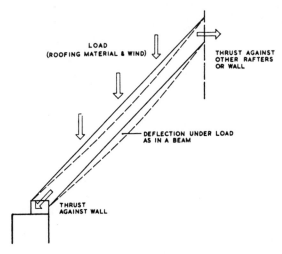

Diagram 91   Function of a rafter

comes on a rafter, slight deflection will take place as shown in diagram 91, so that a rafter must have some of the properties of a timber beam, as shown in chapter 4. On the other hand, if the pitch is very steep, obviously most of the load will be passed direct down the member to the plate and it will act more as a strut. This must be borne in mind in settling the sizes for rafter members, and in cases of doubt the strength of the proposed members should be checked by calculation, but for pitches between 30 degrees and 45 degrees for tiles or slates and with rafters 50 mm wide and spaced 400 mm apart, a good rough rule is that the depth of the rafter should not be less than $\frac{1}{24}$th of the span.

Single roof systems
The construction of the roof is regulated by the width the roof has to span and other factors. A couple roof, shown in diagram 92, has rafters thrusting against external walls, which require buttressing at intervals to resist the side-stress. The rafters have no intermediate support, so that if they are not to exceed an economical size, such as 150 × 50 mm, a roof of this kind will not span more than say 5 m. The side-thrust can be avoided if tie-members are placed across the span joining the feet of the rafters, and these tie-members can form the ceiling joists. This is called a close-couple roof. Alternatively most of the side-thrust can be adequately taken if the tie-members are joined to the rafters a short distance from the bottom. This is useful in saving timber and is called a collar roof.

114 · BUILDING TECHNIQUES

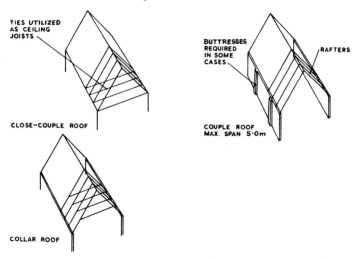

Diagram 92   Single roof systems

Purlin roof systems

For spans much over 5 m some support has to be given in the middle of the building. This can be done with horizontal wood beams, called purlins, supporting the centres of the rafters, and these beams in turn can be strutted in many ways. Diagram 93, 94 and 97 show them strutted from a centre partition, from a partition not in the centre and from cross-partitions. In every case it must be remembered that the purlin is a beam and that it has a load distributed along its length and has to span from one strut to another. It is not always easy to provide these struts at frequent intervals, and if they are wide apart the purlins have to be correspondingly larger. A 225 x 75 mm purlin is necessary to span 4 m between struts. Where support is only given at wide intervals, small purlins can be supported with inclined members (trussed) as shown in diagram 98. These forms of purlin roof systems are known as double roofs.

In some cases where other support is not possible or where the roof space requires a strong floor for storage, the purlin can be strutted from that floor, as shown in diagram 95, by laying a plate across the floor joists and strutting from this. These struts can be close together; in fact for bedrooms in roofs a stud partition is sometimes used and the plate at the top of the partition then becomes the purlin.

An ingenious roof system is also shown in diagram 96, where the struts supporting the purlin are carried on a binder which is hung from

ROOFS · 115

Diagram 93   Purlin roof with centre partition

Diagram 94   Purlin roof with offset partition

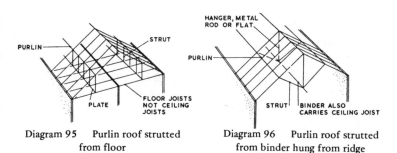

Diagram 95   Purlin roof strutted from floor

Diagram 96   Purlin roof strutted from binder hung from ridge

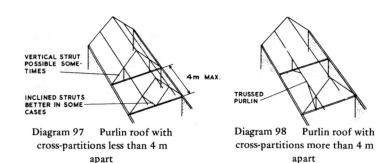

Diagram 97   Purlin roof with cross-partitions less than 4 m apart

Diagram 98   Purlin roof with cross-partitions more than 4 m apart

the ridge by metal rods or straps. The load is then put back into the rafters, but as a direct thrust down the slope. Timber members can take such a thrust in compression much better than in bending.

Trussed rafter
A system of framing up the rafters by means of struts and hangers is now widely used in house construction. Every fourth or fifth rafter is

116 · BUILDING TECHNIQUES

framed to carry the purlin which, because of the short span, can be an economic section. Details of this form of construction are shown in diagram 103.

Roof truss systems

Where there is no intermediate support for roof members or where the spans are excessive, truss systems are used. There are many traditional and modern methods and the principles of two systems are shown in diagram 99. The kingpost truss is an old type of simple frame which can take one purlin in each slope. Where a roof covering such as corrugated asbestos is used which requires a large number of purlins at close spacing, the second truss shown in this diagram would be better (as shown in detail in diagram 109).

It cannot be too strongly emphasized that when trusses have to be used quite a different principle of construction is introduced. Trusses are separate independent frames which collect a large load and transmit it to points, either on to piers to a wall or on to stanchions, and this immediately suggests a building arranged with several equal bays, such as a factory or hall.

Details of purlin roof construction

Diagram 100 shows the layout of a roof where a purlin system is necessary. If plain tiles are to be used, a pitch exceeding 40 degrees is advisable, as shown in the section, diagram 101. The length of rafter for this span would then be about 5 m from plate to ridge. One purlin

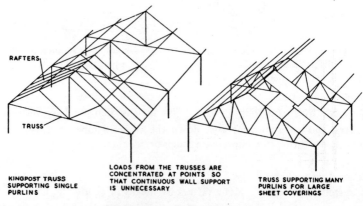

Diagram 99   Roof truss systems

would be required and 160 × 50 mm rafters would be suitable at 400 to 450 mm centres, following the rule given previously.

When laying out a roof, it is advisable to draw a plan to scale and show the position of the purlin and the structural walls so that the support of the purlin from these walls can be determined. In diagram 100 this has been done and it is possible to see that inclined struts, as shown on the section, diagram 101, can be fixed at A. To support the purlin at the hipped end a vertical strut could be placed from the wall at B and C, but the unsupported length of purlin would then be considerable and a large size would be necessary. If two inclined struts are fixed at B, one can support the purlin under the hip rafter and the other can support the centre of the purlin at the hipped end. A similar inclined strut at C will reduce the span of the purlin from the wall to the hip rafter. The greatest length of unsupported purlin in this diagram would only be about 3 m so that a 200 × 75 mm purlin should be adequate. Reference should be made to the tables in Schedule 6 of the 1974 Second Amendment to the Building Regulations for appropriate sites of rafters and purlins for similar types of construction.

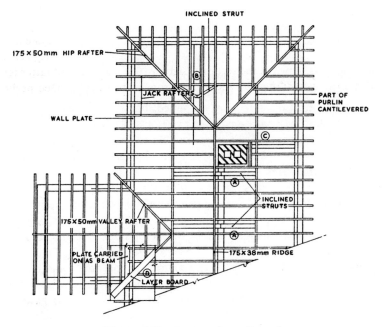

Diagram 100   Layout of a purlin roof

118 · BUILDING TECHNIQUES

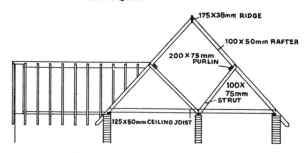

Diagram 101   Section of a purlin roof

Struts up to 2·5 m can be 100 x 100 mm, or 100 x 75 mm if they only take a small amount of load. Struts smaller than this are not sufficiently rigid even if they are short.

Note that the ceiling joists form an effective tie against the spreading of the roof. They would, of course, be nailed to the sides of the rafters and to the plate, so their spacing would be the same as that of the rafters. They need not be as large as floor joists; it is usually sufficient if their depth is $\frac{1}{24}$ th of the span.

To fix the positions for rafters it is necessary to work from fixed points, such as a stack, verge or junction of hip rafters. For fire resistance, it is necessary that the woodwork should be kept away from the stack unless it is rendered. It is therefore advisable to place rafters each side of the stack about 25 mm or so away. Rafters have to be placed near gable walls as shown in diagram 100. It is more convenient to arrange rafters at the junction of the hip rafter as shown in diagram 100, as the erection is thereby simplified, but if this junction comes near a stack it may not be possible to get a reasonable spacing and to have this arrangement at the same time. When these fixed positions are decided upon, the other rafters can be filled in between to give a suitable spacing. Jack rafters should be arranged to coincide on each side of the hip or valley rafter. Where the ends of the rafters are exposed, it may be necessary to make considerable adjustments in their arrangement, as a regular spacing at the eaves is very important. Additional members may have to be fixed each side of stacks, dormers, etc.

The ridge board should be deep enough to support the whole length of the splay cut at the head of the rafter, as shown in diagram 104. For this, a thickness of 38 mm is sufficient. Hip and valley rafters have to be deep for the same reason and in the case of the valley rafter, because in theory it takes an additional load, as it acts as a trimmer for some of the jack rafters.

In some cases it is possible to have stronger hip rafters in order to support the ends of purlins. In these cases the purlin will have to be fixed up to the underside of the hip rafter with bolts, and as the hip rafter will take a considerable load it will have to be increased in size to 225 x 75 mm or 275 x 75 mm. The additional load placed on the wall must be considered and a dragon beam with an angle tie, as shown in some textbooks, is advisable.

In diagram 100 a gable roof over an extension is shown which has the ridge at the level of the purlin. This arrangement is convenient, as the purlin can give support to the ridge when it is fixed and to the ends of the valley rafters.

Small interesting roofs such as dormer roofs can more conveniently be formed without using valley rafters. The main roof would be completed, including the boarding, and the jack rafters for the small roof would be nailed direct to the boarding. If boarding is not used, a layer board as shown at D in diagram 100 would be fixed across the common rafters to support the ends of the jack rafters.

Struts if inclined can conveniently be birdsmouth-jointed to the plates (diagram 102). If it is possible struts should give purlins a small amount of dead bearing as shown in the diagram, where a small notch would not reduce the strength of the strut. Nailing the strut to the side of a rafter helps to stiffen the roof.

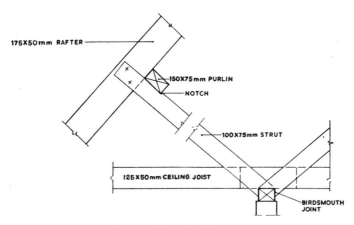

Diagram 102   Roof framing

Trussed rafter detail

A detail of the trussed rafter method of roof-framing mentioned on page 115 is shown in diagram 103. A typical construction suitable for a roof of 7·2 m span, to be covered with interlocking tiles of the type depicted in diagram 89, is shown here. The strength of the framing depends on the strength of the joints and double-sided, metal toothplate connections with bolts and washers, similar to those seen in diagram 110 are used in the construction. It is essential to ensure that these are of the correct make and properly tightened up. The truss shown is of a type copyright by the Timber Research and Development Association and it is important when using this type of roof to remember that the framing is designed to fulfil specific conditions only, and that unauthorized variations in the detailing or in the specification may make the roof unsafe. Note that the design is based on the use of timber graded in accordance with B.S. 4978.

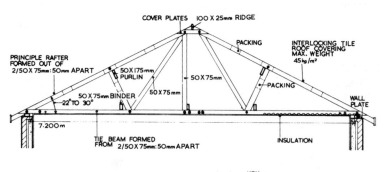

Diagram 103   Trussed rafter detail

Framing and covering details

Details showing parts of typical roof-framing in conjunction with the appropriate covering are shown in diagrams 104 and 105. Diagram 104 shows a ridge detail where insulation board is fixed to the rafters and covered with felt. Laths are used as counter-battens to keep the main battens clear of the felt. Battens for tiling are usually 50 x 25 mm. A shorter tile often has to be used at the ridge, as the battens have to be more closely spaced together, but the margin between the tails of the tiles has to be preserved. Hog-back ridge tiles as shown give a good appearance and can usually be arranged so that they cover the nail hole

ROOFS · 121

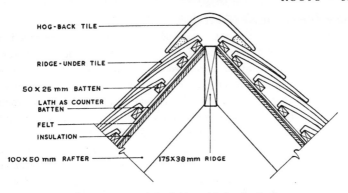

Diagram 104   Detail of ridge with hog-back tile

of the second course of tiles. They are usually about 600 mm long and should be bedded in gauged mortar only at their edges and at their junctions, where they are pointed up also in gauged mortar. It is unsound to bed them solid, as the timber would be so covered that it might rot.

Other forms of ridge tile are used, some having overlapping joints, which are more efficient but unsightly.

Diagram 105 shows an eaves detail for a plain tile roof where the rafter has been carried on to give a projection of about 150 mm. A

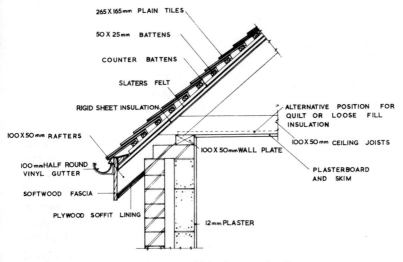

Diagram 105   Plain-tile eaves detail

150 × 25 mm fascia board is fixed vertically to the ends of the rafters and a soffit is formed with external quality plywood screwed to the rafter. A deep fascia is necessary as the top supports the bottom course of tiles; the felt covering is dressed over the top of the fascia and the counter-battens have to be kept back from the eaves so as not to foul this change in plane. Note the 'under-eaves' tile, which is shorter than the normal tile. Note also the provision of a layer of insulation on top of the rafters. The insulation can be in sheet form, but quilt or loose-fill insulation could also be provided as an alternative in the spaces in between the ceiling joists as indicated in order to obtain the U value required in the Building Regulations.

The finish of plain tiling at hipped ends is usually done by using special hip tiles, one kind being shown in diagram 106. They are fixed by a single nail to the hip rafter and pointed up in gauged mortar. Tiles have to be cut to fit against bonnet hips and so a number of 'tile and a half' tiles have to be used. These are shown and explained in connection with verges later.

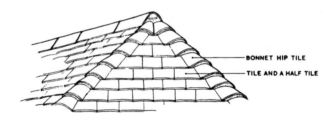

Diagram 106   Bonnet hip tiles

Hog-back ridge tiles or half-round ridge tiles can be used to cover hips, but they are large and unsightly. Special hip tiles giving a sharp angle are effective but are also not so pleasant in appearance as bonnet hips.

The intersection of tiling at a valley can be done with special tiles of reverse section to those used on the hips, or by forming traditionally an open lead gutter, or alternatively by using lead soakers, as shown in diagram 107. These are made from no. 5 lead and are the length of the covered part of a tile plus 50 mm for fixing. In this way the side joints between tiles are always covered but the lead soakers never show. A preformed fibreglass section gutter can now be used as an alternative to the traditional open lead gutter construction.

Verges can be finished as shown in diagram 108 by bedding a tile in

ROOFS · 123

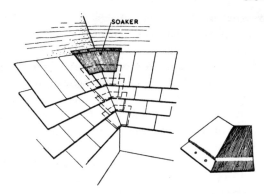

Diagram 107   Valley gutter with soakers

mortar on the top of the gable wall. The battens are laid across the wall and bed on this tile, being cut back about 50 mm from the outer edge. The tiling will be hung on the battens in the normal way with the outer edge coinciding with the outer edge of the 'underverge' tile. It will be seen from the diagram that, to keep the bond, wider tiles are needed in alternate courses. These are called 'tile and a half' tiles and are a normal 'special' for all tiles that are made. The edge of the tiling is then pointed up with gauged mortar.

Other finishes are sometimes used, particularly where a wide projection is required. Then a rafter has to be fixed outside the wall and will be wrot and painted or cased in with a barge board and soffit board. In this case the rafter has to be supported by projecting ridge, purlin and plate, and if the projection is wide, these have to be supported on brackets.

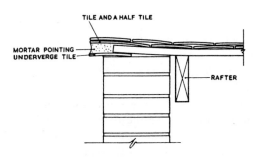

Diagram 108   Section of verge construction

Laminated truss detail

The traditional method of spanning 6 m to 9 m using timber was to frame up a kingpost roof truss, the outline of which is shown in diagram 99. This construction relied on large sections of timbers and is now obsolete. A laminated truss shown in diagram 109 would now be used as an alternative.

This is formed with members built up of two or three thicknesses of boards. By this method the members can be arranged to intersect at a point which avoids bending stress in the timber. The members can be quickly and easily made by nailing the boards together or using ring or toothed connectors (diagram 110) and complicated carpenter's joints as required for the kingpost truss are avoided.

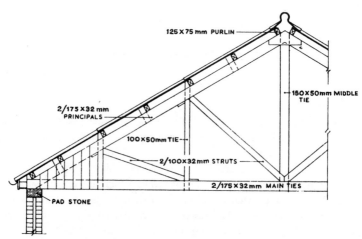

Diagram 109   Laminated truss

Flashings

Joints between slates or tiles and vertical parts of a building which intersect the roof, such as parapets, chimney stacks, dormers, etc., have to be waterproofed, and this is best done with sheet metal, such as lead, copper or zinc. In cheap roofing cement fillets can be formed, but they are inclined to crack.

Diagram 111 shows a chimney stack where most of the flashing details are incorporated. Where the flashing is required at the top or above the tiles, an apron can be formed as shown on the front of the stack. This should be tucked into a convenient joint in the brickwork, wedged and pointed with mortar. The vertical face should be not less

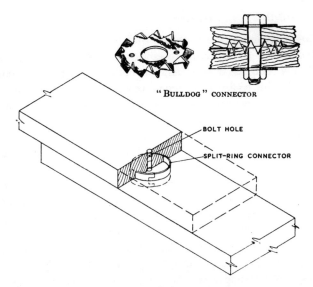

Diagram 110   Timber connectors

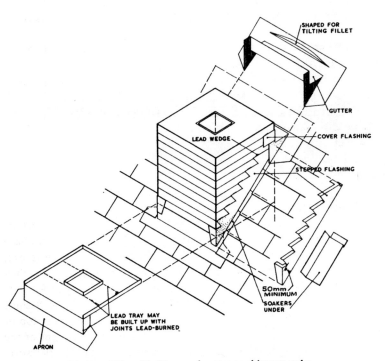

Diagram 111   Flashings and gutter to chimney stake

than 150 mm. The lower part should then be dressed over the slates or tiles for 150 mm, or sufficient to cover the nail holes of the second course below. The raking joint between a roof covering and the stack is best formed with lead soakers similar to those used in a valley. These soakers will be square as shown, the side being turned up 75 mm. A cover flashing is fixed to the brickwork to cover these upturned edges, and this has to be cut in a series of steps with the horizontal edges turned in to the brickwork, wedged and pointed. Where a great length of stepped flashing is required, joints in the length have to be formed by lapping the higher over the lower for 100 mm.

Behind the stack a gutter is necessary and this should be sufficiently wide to allow a tiler repairing the roof to stand without breaking the edge of the tiles. The horizontal part of the gutter has to be formed with boards on bearers fixed to the rafters or trimming member, and the lead gutter can, in the case of small stacks, etc., be formed in one piece as shown. The upper edge should carry some distance up the slope of the roof and be tucked over a tilting fillet, which is necessary to lift the edge of the tiles as in the eaves detail.

It is important that a damp course be inserted in chimney stacks. This can be done with bituminous felt or slates. The damp course should be at the level of the top of the apron at the front of the stack. This means that an area of damp brickwork is exposed in the roof space. In large enclosed roofs this may not matter, as evaporation by ventilation will prevent the moisture spreading to surrounding woodwork, but where this damp brickwork may cause damage it may be advisable to drain this part of the stack by forming a lead tray, as shown in diagram 111.

CHAPTER SEVEN
Doors

Timber
Timber used in building was discussed in chapter 5. The various species of wood were there set out and the different methods of conversion for different purposes were mentioned.

For joinery it is essential that the timber is seasoned to a moisture content to match the humidity of the room in which the work is to be fixed. Some swelling and shrinkage may still, however, take place, as, in practice, it is often found difficult to prevent woodwork from coming into contact with wet building materials during the building process.

For these reasons timber should be very carefully chosen to be reasonably free from knots, to have a straight grain and to be of a correct cut from the log so that the deformation in use of each wood member is kept to a minimum.

As described in chapter 5, the maximum shrinkage is across the grain; shrinkage in the length of most timber members is not enough to worry about. The width of individual boards, therefore, should be kept as small as possible and joints arranged by overlapping or tonguing so that shrinkage does not lead to unsightly cracks across plain surfaces. Shrinkage cracks which appear on internal angles are not so evident. The fixing of timber members must be such as to allow this shrinkage to take place. For instance, if two nails are placed opposite each other near the outside edges of the battens in the door in diagram 112 splitting on the line of one nail or the other is sure to take place. One nail can be used near one edge, the other edge being held in position by the tongued and grooved joint.

Hardwoods are much used for doors and windows where a very good finish is required. They are stronger, more durable and usually permit members of a smaller size to be used.

It is important in detailing joinery to remember that it is economical to select timber of standard size, i.e. 19 mm, 25 mm, 32 mm and 38 mm thick, etc., but when these timbers are planed they are reduced by approximately 1·5 to 3 mm for each surface; therefore 50 mm doors usually finish at 44 mm thick or slightly over.

Doors

Diagrams 112 to 119 show the most common range of doors. Diagrams 112 and 113 show ledged doors, which are economical but only suitable where appearance is not important. The ledged and braced door in diagram 112 consists of 25 mm battens running vertically tongued and grooved and held together by the ledges. The door is hung to the frame with tee hinges, sometimes referred to as cross-garnets. As the centre of gravity of the door is some distance from the hinge there is a tendency for the lock edge to drop, so braces are cut in as shown. Note the arrangement of the joints which take the direct thrust of the weight down to the hinged side.

Diagram 113 shows an improvement on diagram 112 in that the door has a frame of 100 mm stiles and top rail and 225 mm bottom rail, framed together with mortice and tenon joints, the infilling only being of battens with diagonal braces as before. This makes a very strong door much used for agricultural buildings.

Diagrams 114, 115, 116 and 117 show framed panel doors which will now be found in older buildings. These doors usually have stiles, rails, etc., out of timber 50 mm thick. For doors much larger than the sizes shown the thickness should be increased to 63 mm. The principle of the construction is shown in diagram 120. The tenons are on the rails and should pass right through the stiles. They must, of course, be cut back or haunched so that the mortice is a sufficient distance away from the end of the stile. If the tenons take up a third of the thickness of the stile, and if they are well wedged, the joint can be very strong. This is essential in these doors, as the joint has to resist all tendency for the lock edge to drop.

In the single-panel door, diagram 115, the bottom rail is kept deep so that the tenons can be large and strong. In the four-panel door, diagram 114, an additional rail, the lock rail, is introduced with similar tenons. This increases the strength and enables the bottom rail to be reduced in depth.

Between the stiles and rails, panels are formed which traditionally were generally filled with boards about 15 mm thick. In single-panel doors the panel could not be out of one board, and so would have to have joints which would be formed with concealed tongues and glued. The complete panels would fit into grooves in the stiles and rails (see diagram 120), but would not be fixed rigidly. The wide bolection

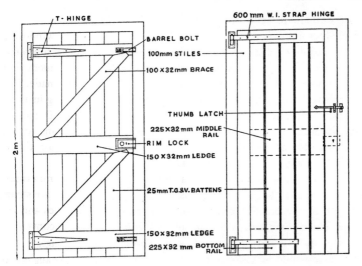

Diagram 112 Ledged, braced and battened door

Diagram 113 Framed, ledged and braced door

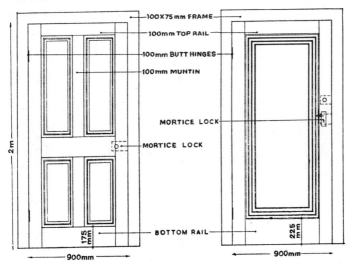

Diagram 114 Four-panel door

Diagram 115 Single-panel bolection moulded door

mould (diagram 122) would cover these joints and allow for the expansion and contraction of the panels. In doors with lock rails and muntins the panels would, of course, be much smaller and in some cases cross-tonguing would not be necessary and the risk of splitting reduced.

It is common practice now, however, to form the panels in plywood, which does not expand or contract.

These doors are shown hung on 100 mm butt hinges. Steel butts are commonly used, but brass butts have a longer life, particularly if they have nylon bearing surfaces. It is more common and gives greater security to have mortice locks for these doors. These locks are very thin and can be inserted in a large mortice in the edge of the door, so that they cannot be removed when the door is shut as the rim locks (diagram 112) can be. Double or twin tenons are used in thicker doors, which will allow for the fitting of the mortice lock.

Diagram 116 shows a four-panel door as adopted in B.S. 459. This is an economical door, but with three cross-rails there is sufficient strength in the mortice and tenon joints to ensure that it retains its shape. This door is usually 35 mm finished thickness.

Diagram 117 shows a glazed door. The glazing bars should be as small as possible in cross section: say 44 x 22 mm to give an elegant profile. They do not add to the strength of the door but only divide up the area into small squares which permit thin glass to be used. The glass should be fixed with wood beads, as shown in diagram 126, the edge of the glass can be fitted into a synthetic rubber or vinyl gasket.

Diagrams 118 and 119 show two types of flush door. Both surfaces of these are covered with plywood glued on, which has considerable

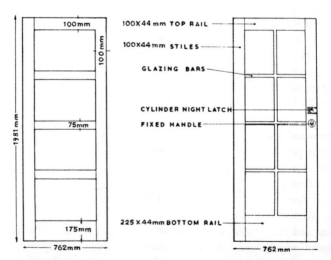

Diagram 116   Four-panel standard door

Diagram 117   Glazed door

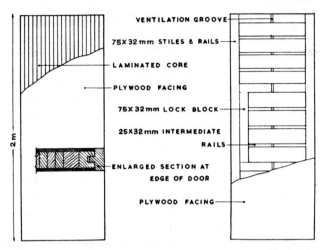

Diagram 118 Flush door: laminated core

Diagram 119 Flush door: framed core

strength and will not deform, so that framing with mortice and tenon joints is not so essential. Diagram 118 shows a solid door with a core made up of strips of softwood with the plywood covering glued on each side. If the grain of the softwood strips is arranged so that the annual rings are in alternate directions, the warping of one strip is counteracted by the warping of the next strip.

Diagram 119 shows a flush door which uses less timber, in that the ply sheets are stuck on to a light framing. Blocks are fixed in this framing to give a solid area where mortice locks may have to be fixed. It is important that the hollow spaces should be ventilated, so grooves are usually formed in the members as shown.

Fire-resistant doors

In certain positions in a building, doors must be able to resist the spread of fire for a given period, and it is possible by incorporating fire-resistant sheets in the design to enable a flush door to be used where a 'fire-check' period of up to 1 hour is required. This is done by fixing plasterboard sheets on either side of the door between the framing and then flushing the door on each side in 5 mm thick asbestos wallboard which is in turn faced with plywood as a decorative finish. This type of door is very heavy and should be hung in a hardwood (or fire-protected softwood) frame with deep rebates. If the door is only

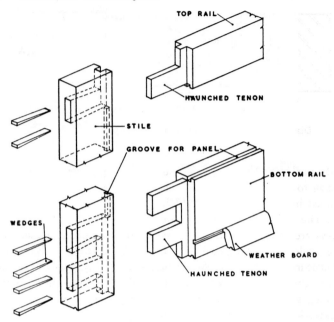

Diagram 120   Mortice and tenon joints for stiles and rails

required to have a fire resistance of half an hour, the asbestos wallboard is omitted. Flush fire-check doors are fully described in B.S. 459, Part 3.

Sound insulation doors
Better standards of building demand better sound insulation, and an inexpensive lightweight internal door may not be adequate in this respect, since a flush door of this pattern would only give a sound reduction of 15 dB. This means that sound caused by the rustling of paper or whispering would be heard through the door. By fitting draught-excluders and using a solid-core door of the type shown in diagram 118 the sound reduction can be improved, but to prevent conversation being overheard, particularly if the voices are raised in argument, then two closely fitting doors separated by an airspace must be used.

Door frames and linings
Doors can either be hung to frames, which are strong members framed together, or to thin linings to reveals which can be made strong enough to support a door.

DOORS · 133

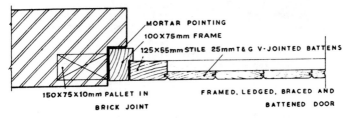

Diagram 121   Plan of framed, ledged and braced door and frame

Diagram 121 shows a frame out of 100 x 75 mm which is strong enough to take a heavy framed, ledged and braced door. The posts are tenoned into the head and sometimes dowelled as shown in diagram 123. The bottom of the posts should be dowelled to the floor. These frames are usually built in, as they are strong enough to take heavy wear while the building work is going on. Fixing to the walls can be by wrought-iron straps, as shown in diagram 113, or they can be nailed to lightweight concrete blocks, fixed to wood pallets built into a joint, or to wedges or plugs driven into joints. Wedges are liable to move the brickwork. Pallets are usually too thin to give good fixing.

For external doors, frames are also used as shown in diagram 122, the exposed angles being moulded. Jointing and fixing are as before described. It is important that the gap between the frame and the wall should be filled. Jointing with gauged mortar is the most common method, but this is not draughtproof and it is better to fill this joint with mastic driven in with a pressure gun. This also shows a wood lining to the inside reveal to the door opening, tongued to the frame and fixed to a rough ground which also serves as fixing for a wide moulded architrave. The lining is free to shrink if it is fixed only to the rough ground, as the tongue at the other edge can draw slightly out of the groove without any harm being done.

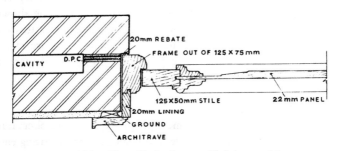

Diagram 122   Plan of bolection moulded door and frame

134 · BUILDING TECHNIQUES

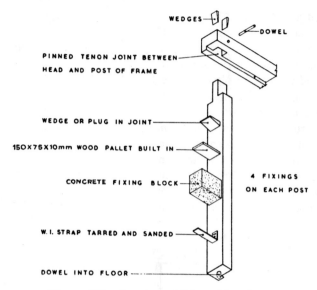

Diagram 123   Framing and fixing of door frame

Diagram 125 shows a typical lining to an opening in a half-brick wall. Cross-pieces can be fixed to the brickwork on the reveals and adjusted so that the faces are in a true plane to take 140 x 22 mm lining. The jamb lining cannot be tenoned to the soffit lining, so a

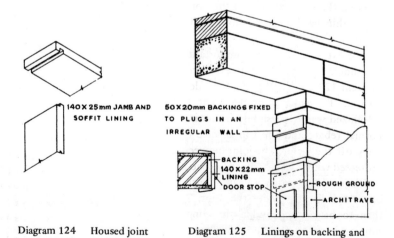

Diagram 124   Housed joint between jamb and soffit lining

Diagram 125   Linings on backing and rough grounds

housed joint as shown in diagram 124 is usual. The rough grounds give additional fixing for the lining and support the architrave. The rebate for the door is formed by a 75 x 13 mm door stop. The 22 mm lining will not support a heavy door, because there is not sufficient thickness of wood to hold the screws on the hinges. In some cases this is got over by having three hinges instead of two, or fixing blocks behind the lining where the hinges occur.

For thin partitions it is sometimes convenient to make the width of frame fit the total thickness of the partition, as shown in diagram 126. Here, if the partition is of strong material such as lightweight concrete blocks, a 32 mm thick lining would be sufficient. The architrave, a simple flat fillet with rounded edges, is moved across on one side to cover the joint between the stop and the lining.

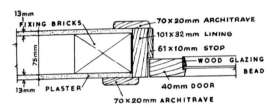

Diagram 126   Door lining in block partition

The bottom of internal doors usually presents no problem, but where a thick carpet is used the bottom of the door has to be kept up and a thin hardwood sill sloped off each side can be fixed under the door.

For external doors it is important that a closer fit is obtained at the threshold. It is good practice to provide a hardwood sill as shown in diagram 127, which is part of the door frame, the jambs being tenoned into it. To make a proper joint below the sill a wrought-iron or galvanized water bar should be set in the threshold. The door is set in a normal rebate and the top surface of the sill should be weathered to carry off the water. The joint between the door and the sill can then be protected with a weatherboard as shown.

Diagram 128 shows a neater finish with a stone sill in which a water bar is set directly under the door. The bottom edge of the door has then to be rebated and a throating should be cut to prevent drips running back on the under surface. This method is more difficult to construct efficiently. Another detail is shown in diagram 35.

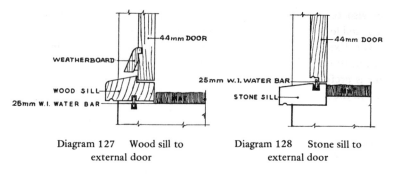

Diagram 127 Wood sill to external door

Diagram 128 Stone sill to external door

Door furniture
Latches
The door is not complete unless provided with a handle, lock, latch or other means of fastening. This is known as ironmongery or 'door furniture'. There is a very large range of furniture from which to choose but the basic requirements for each individual door are usually relatively simple. If the door is only to be held closed and is not required to lock, a latch will be sufficient. This could be either a 'rim' latch, fitted on the surface of the door, or a 'mortice' latch fitted within the thickness of the door. The latch will be actuated by a spindle turned either by a lever handle or by a knob. The latches will not necessarily be obtained from the same manufacturer who supplies the knobs, and care should be taken to ensure that there is enough clearance (minimum 55 mm) from the edge of the door to the spindle so that the knob may be turned without damaging the knuckles on the door lining.

If British Standard doors are used, the lock block in a flush door allows the furniture to be fixed between 600 mm and 900 mm from the floor.

Locks
If the door is to be secured, the basic choice is between a 'rim' lock and a mortice lock, with the option of a 'night-latch' action which is an inside fixed-rim lock operated by a small key from the outside and a knob from the inside. There are many variations of this type of lock with special burglarproof devices.

It is very likely, however, that the most convenient arrangement would be door furniture which allows a door not only to be held closed but also to be locked. Thus the two-bolt lock, which combines both latch and lock mechanism, is probably most used.

The fixing plate behind the lever handle or knob must be large enough to clear the lock case and allow long fixing screws to be used. This is preferable to a small circular plate (or rose) since English locks rely on the pull of the door being taken by the plates.

CHAPTER EIGHT
Windows

There are many types of window. Some are complicated, and are usually for particular purposes of ventilation, lighting or view, to suit the requirements of schools, hospitals and special buildings. Nearly all windows, however, consist of two parts, a frame, which can be built or fitted into the opening in the wall, and sashes (or casements), which are smaller frames containing the glass, attached to the main frames by hinges, or retained by grooves. It is the way the sashes are attached which makes the chief difference and which gives so many varying possibilities for different uses.

Principles of framing
Diagram 129 shows the principles of framing applied to a side-hung casement window. Window frames should never be used to support the brickwork over the opening. When this has been done in the past there are many cases in which settlement in the brickwork has now deformed the frame, making opening and closing of the casement very difficult. In extreme instances movement in the brickwork is enough to crack the glass in the frame. The brickwork over the window should always be properly supported on a lintel.

The frame can be secured by means of metal lugs and by building in as the work proceeds or, alternatively, by first building plugs into the brickwork and then securing the frames after the opening is formed. Frames should be bedded in a weak cement/lime mortar, and afterwards pointed upon the outside by mastic. The mastic is usually applied by a pressure gun. The joint between the jamb and head of the window will be a through tenon similar to that used in a comparable position in the construction of a door frame and illustrated in diagram 123, chapter 7. Note, however, that it would be impracticable to wedge the joint due to the comparatively small cross section of the timber, unless the mortice was tapered. The joint would thus be glued and secured by a hardwood dowel or a metal 'star' dowel. The dowel is used

WINDOWS · 139

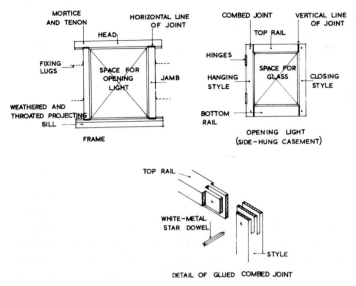

Diagram 129   Principles of framing a window

to fix the joint while the glue sets. With this type of joint projecting horns are provided which will be built into the brickwork.

An alternative to the tenon joint is a combed joint which would also be used for the corners of the casements. This type of joint, which is now universally used in factory-produced joinery, has been specially developed for machine manufacture and relies for its permanence on the strength and durability of the weather-resisting glue that is used to secure it. The joint gives proportionally greater gluing area than the traditional mortice and tenon. Note from the diagram that the strength of the side-hung opening light which functions as a 'door' relies on the depth of the bottom rail.

Window analysis

To construct a first-class window with an opening light is not an easy matter since the casement must fit tight against the frame to make a draughtproof joint and yet not bind against it, making opening and closing difficult. The casement must be strong enough to withstand continuous opening and closing without pulling out of shape and yet at the same time the timber sections must be elegant so as to let in as much daylight as possible. Diagram 130 analyses the profile of a double-rebated casement. The sections shown are based on those

140 · BUILDING TECHNIQUES

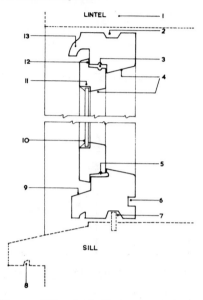

Diagram 130   Analysis of casement window

(1) Brickwork at head supported on lintel.
(2) Groove to act as key for mortar when setting frame in opening.
(3) Twin capillary grooves and chamfer to prevent water gaining access by capillary attraction.
(4) Reveals chamfered to reflect light into the interior and to reduce width of timber seen against light.
(5) Groove and chamfer to prevent driving rain being blown under casement.
(6) Groove to receive windowboard.
(7) Groove to locate weatherbar.
(8) Throat to capture water blown under sill and allow it to drip clear of the face of the building.
(9) 'Weathered' sill; sloping surface to take away water which runs down face of glass.
(10) Glass cut 2 mm less than opening.
(11) Rebate to receive glazing.
(12) This joint should be as close a fit as possible.
(13) 'Drip' to prevent rain driving down behind opening light.

recommended by the British Woodwork Manufacturers' Association and used principally in housing. Joinery that is made to the designs and standards of the Association is marketed under the trade name 'EJMA' and is produced under licence from the Association.

Opening lights

Diagram 131 illustrates some alternative methods of arranging the opening sash or casement within the frame.

WINDOWS · 141

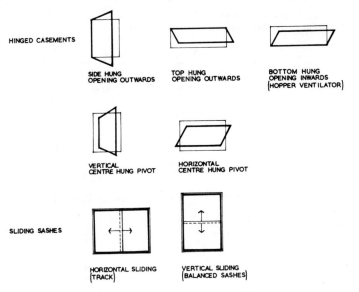

Diagram 131   Types of opening lights

Timber windows

Diagram 132 shows a typical rebated timber casement window.

The size of the members depends upon the size of the window opening and quality of wood obtainable, hence no dimensions are shown.

Metal windows

The most usual types of metal window are either rust-proof steel, or aluminium, preferably with weather-stripping. Steel windows are available from the following ranges:

(1) The standard range intended mainly for domestic use (B.S. 990) Module 100 windows. The windows of 600, 1200, 1500 and 1800 mm wide and 200, 500, 700, 1100, 1300, and 1500 mm high incorporate side- and top-hung casements opening outwards together with a range of glazed doors 2100 mm in height. Diagram 133 shows this type of window. The units can be coupled together using standard fixed lights with the basic opening lights to form composite glazed wall units as shown. The glass can be secured by small wire clips and glazing compound, or metal glazing beads with special screw fixings can be used. Note that where glazing compound is used it must not spread beyond the line of the frame to obscure the clean line of sight.

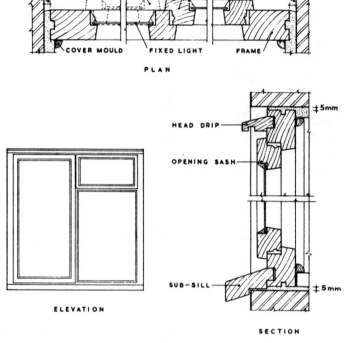

Diagram 132   Double-rebated wood casement window

(2) Windows formed of heavier-duty rolled-steel sections known as W20 metric range. This produces windows similar to the Module 100 range but without the traditional small top-hung 'night-ventilator' pane. The W20 section is also used for windows to special sizes purpose — made for a particular job.

(3) A separate Standard intended for industrial buildings and capable of being linked together in composite form to provide large glazed areas (B.S. 1787) in W20 sections to special sizes.

(4) A limited range of standard windows intended for use on farm buildings. This range has comparatively small panes of glass and has the inward-opening hopper ventilator as standard construction, to provide simple draught-free ventilation (B.S. 2503).

Standard ranges of metal windows can also be obtained in aluminium but, in addition, horizontal and vertical sliding windows, which are difficult to make in steel, can be produced very elegantly in the lighter metal. Aluminium windows are supplied natural finish or anodized. The natural finish will soon become dull and pitted in a polluted atmosphere unless regularly maintained by washing.

Window surrounds
Metal windows can either be built direct into the opening against the brickwork by means of an adjustable fixing, as shown in diagram 133, or they can be supplied with standard metal or timber surrounds. The disadvantages of 'building in' without surrounds, as the brickwork proceeds, are the difficulty of maintaining a sufficient tolerance gap all round the frame and the risk of damaging and distorting the window while building up.

Where standard wooden surrounds are used, they should be primed

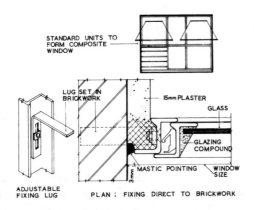
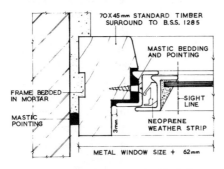

Diagram 133   Modular metal windows

and undercoated on all surfaces before fixing. The steel window is bedded in mastic and secured by zinc-coated screws to the subframe, and then both frame and surround can be fixed at any convenient time after the walls have been built up and the openings properly made. This method of fixing the window is more expensive than building in as the work proceeds, but it has the additional advantage that the frames are less likely to be damaged by the other building operations, or by stresses transferred from a badly built opening. A standard timber surround to B.S. 1285 is shown in diagram 133.

Metal surrounds which can be made from say 10 gauge (3 mm thick) pressed steel, save time on the job, as they can be built in and are strong enough to withstand heavy wear, simplify plumbing of angles and can act as shuttering for short brick lintels. They also form a convenient way of closing the cavity. Typical sections are shown in diagram 134, which shows the standard metal casements fixed to the outer angle with screws. The pressed-steel sill should have a proper turn-up at the ends to ensure that water runs off the front edge. A satisfactory finish can be obtained by turning the plaster into a recess formed in the surrounds, as the surrounds are completely rigid.

Sliding sash window

The traditional timber sliding sash window in cased box-like frames which relied on heavy weights to counterbalance the sashes by a system

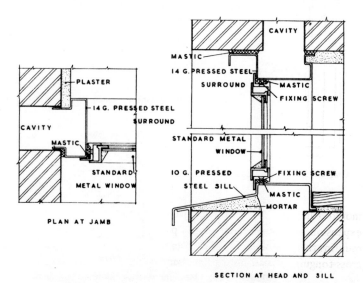

Diagram 134   Standard metal window with steel subframe

WINDOWS · 145

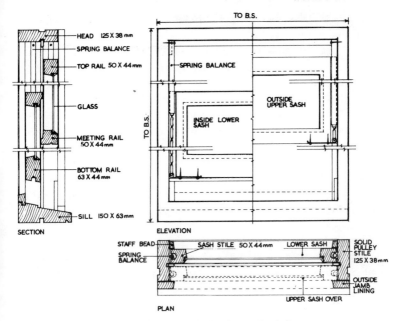

Diagram 135   Standard sliding sash window

of cords and pulleys is now becoming obsolete because of the expense of manufacture, the bulky detailing and the cost and time required for maintenance. It has now largely been superseded by a balanced sash controlled by a sliding spring mechanism as detailed in diagram 135. The sash window when well made is an excellent means of construction since it allows unobstructed vision and at the same time gives very efficient and easily controlled ventilation. The window detailed is a standard type to B.S. 644 (Part 2) and would be supplied with the sashes already hung. Note that the sash mechanism can be adjusted or replaced very easily.

Glazing

Clear glass for glazing is manufactured by the float process and is made in thicknesses of 3, 4, 5, 6, 10, 12 and 15 mm. Float glass has parallel and flat surfaces which give undistorted vision. The safe thicknesses of glass for a particular window is not only related to the size and proportion of the window (length in relation to width) but also to the wind pressure likely to be experienced on the face of the building.

Glass reinforced with a wire mesh is used where there is a danger of broken glass falling and hurting people. The reinforcement does not strengthen the glass materially, but holds any broken pieces together. It is also good in resisting fire.

Small panes of glass are fixed with putty. A small quantity of putty should be put in the rebate and the glass pressed against this before the face putty is placed and smoothed with a knife. Putties with linseed oil are good for wood sashes, but putties which dry quicker are more suitable for metal windows, although they are difficult to remove when replacement is necessary. Larger sheets of glass should be fixed in with wood beads.

In roof glazing, as used for factories, sheets of wired cast glass 600 mm wide and up to 2·4 m long can be laid very efficiently, resting on patent glazing bars of aluminium or steel, the glass being held by lead or aluminium strips instead of by putty.

Double glazing

In order to keep pace with the rising standards expected in building, double-glazed windows are in greater demand. The objects of double glazing are to cut down the loss of heat and to minimize the transmission of sound through a window. As the name implies, the technique is to provide two panes of glass which enclose an air space. The two panes may be assembled as part of the manufacturing process to produce a 'double-glazed' unit consisting of two sheets of glass separated by a small spacer and enclosing a cell of dehydrated air. The distance between the sheets of glass can vary from 3 to 12 mm, the size of the two units being limited according to the thickness and type of glass chosen. The unit is glazed into a window frame as if it were a single sheet of glass. Alternatively, two frames each separately glazed with a single sheet of glass may be placed in position to give the necessary air space. A third alternative is to fix two sheets of glass in one frame. With the two latter types precautions should be taken to prevent moist air from the inside entering the air space and the air space should be allowed to 'breathe' to the outside air. Fourthly, it is possible to have twin frames each glazed with a double-glazing unit. Diagram 136 compares the techniques. It is common practice to take precautions to prevent the loss of heat through wall and the roof spaces so it is logical to show the same concern in respect of windows which take up an increasing proportion of the total area of the walling. Heat loss through glazed areas can be halved by using double glazing. The thermal transmittance coefficient for a single sheet of glass is $U = 5·7$, and for double glazing with a sealed unit having 6 mm air space the U value is 3·0. A further thought to be borne in mind is that condensation, which

WINDOWS · 147

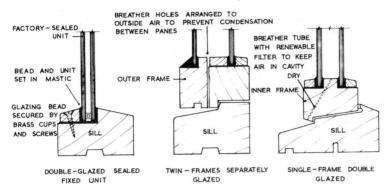

Diagram 136   Double glazing

can be particularly troublesome in display windows, is largely eliminated. If double glazing is used primarily for sound insulation, then the air space must be increased. The maximum effect is obtained by separating the glass by as much as 250 mm while 75 mm should be considered the minimum. This means that a 'twin-frame' unit must be used.

There are many proprietary methods of fixing an extra pane of glass on to existing windows in order to improve the thermal insulation. The extra pane of glass is usually fixed on the inside of the frame or else in the window reveal. It should be made easy to remove the additional pane of glass for cleaning. Diagram 137 shows a horizontally pivoted double-glazed twin-frame window unit of the type developed in Scandinavia. The frames can be obtained in laminated softwood or solid hardwood, or with the outer softwood frame clad in aluminium. The type of window illustrated must be fixed in a preformed opening and should never be built in as the work proceeds. The frame must be tightly wedged at the jambs just below the hinge position so that it does not distort and thus prevent the pivot action. The two frames are normally coupled so that they open as one unit with a special arrangement of hinges so that they can be separated for cleaning and maintenance.

Wall glazing
Very often the technique of glazing in wooden frames is used to construct a floor-to-ceiling 'wall of glass' into which opening lights and solid infil panels are set. A detail of this type of construction is shown in diagram 138. Care must be taken in the orientation of designs

148 · BUILDING TECHNIQUES

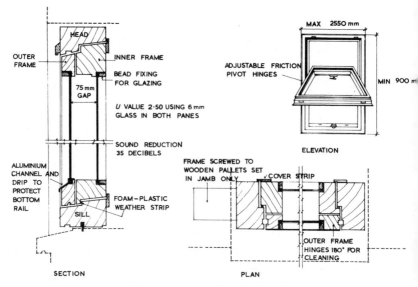

Diagram 137   Double-glazed window

incorporating floor-to-ceiling glazing, otherwise the occupants of the building may suffer discomfort due to excessive solar heat gain during the summer months. Centre-hung horizontal pivot opening lights have been used to give high-level ventilation and panels of opaque glass and boarding have been used below sill level. Note that the cavity behind the boarded panel is ventilated to allow any condensation to escape.

Curtain walling

The use of glazing within a frame to form a floor-to-ceiling external wall panel has already been mentioned and this technique can logically be developed for use in multistory work. The wall then becomes a 'curtain' that is drawn or hung over the façade, and the framework can be of timber, aluminium, steel or even concrete. This type of cladding is known as 'curtain walling' and although the panel is self-supporting it is essentially non-loading-bearing, being clipped back to the façade at convenient points. The curtain wall must be light in weight, weather-resistant and durable, and it must provide adequate thermal insulation. The wall must also give the minimum period of fire resistance specified in the Building Regulations. This is often a difficult matter, particularly where the framing is of timber.

A curtain wall is essentially a series of glass panels jointed to thin

WINDOWS · 149

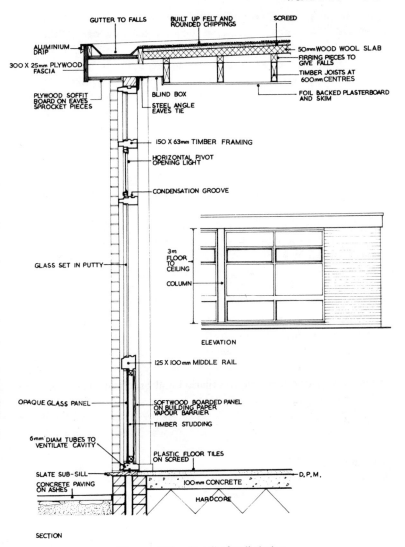

Diagram 138   Detail of wall glazing

structural mullions and transoms. Where non-glazed areas are required, these are filled in with lightweight panels of an impervious outer skin with a core of insulation and a decorative internal finish. The use of impervious materials provides a thinner wall than would be possible with the traditional porous brick or stone.

In the design of a curtain wall, formidable problems of thermal movement and weather tightness at the joints must be solved. Since the façade, is impervious, every drop of rain which falls upon it will run down over the surface and thus any weakness in the jointing will result in a leak to the inside. The thermal movement which takes place in materials under changes of temperature must be studied and allowance must be made in the joint design to accommodate this. Tolerances and a weather-tight seal between different materials, which have always been paramount considerations in building construction, assume overriding importance in curtain walling. The use of light-coloured materials and polished surfaces tends to reduce the problem, but failure to provide adequate tolerances and poor workmanship in fixing will lead to a failure of the construction.

The widespread use of proprietary mastics or synthetic rubber seals is an attempt to provide an impervious joint which at the same time allows adequate movement. A mastic should be capable of maintaining a permanently flexible joint between 'moving' surfaces of different kinds and at the same time provide a lasting seal against moisture or draughts. This is technologically a difficult thing to achieve, in particular since mastic hardens gradually with age and will sooner or later pull away from the sides of the joint which moves. Thus it will be understood that very careful detailing is essential and that the correct type of mastic must be used. The advice of the manufacturer should be sought at an early stage and the claims for the material checked against the job requirements.

The best known synthetic rubber sealing material is 'Neoprene' which is widely used to form a gasket into which glazing is fitted.

System building

Modular units and prefabricated components have always been a tradition in the building industry. Bricks and blocks, and joinery units such as windows and doors have long been produced within a given range of standards.

Industrialized or system building is thus the more complete utilization of the techniques of factory manufacture to the production of an increasing range of building components. These components are then assembled on site to produce a completed building.

It is a method which can take advantage of modern industrial capacity for the prefabrication of repetitive units and which makes closer supervision and more accurate workmanship possible because the components are manufactured under factory conditions. Thus the components must be capable of economic factory prefabrication within certain limits set by the problems of handling and transport.

It will be seen that overall standardization of dimensions is essential for the development of an 'open system' of building in order to make a complete range of modular components available.

With most system building, site works consist of the traditional excavation and foundation techniques up to the ground-floor slab. The factory-made structural frame or complete wall panels are then erected and fixed on the module ready to receive the prefabricated upper floor and roof units. The structural framework is then clad either by traditional materials used in modular form or else by factory-made panels. Doors will probably be delivered already hung between the frames, and windows will be preglazed at the factory. One of the main attributes of system building is the use made of 'dry' techniques of construction.

CHAPTER NINE

Staircases

Staircase layouts

There are many common arrangements for the layout of staircases. The simplest, shown in diagram 139, is a straight flight. If there are more than fifteen steps it is essential to have a landing, and it is often more convenient to have a landing in shorter flights.

Where the 'straight run' space is limited a common form of staircase is the quarter-turn shown in diagram 139, which consists of two flights with a landing between; a person mounting has to turn through a quarter of a circle. This landing is called a quarter-space landing.

Alternative arrangements are the dog-leg and open-well stairs, in which the position one arrives at on the floor above is almost over the position from which one started on the floor below. The dog-leg staircase has a half-space landing, the open-well stair two quarter-space landings. The latter gives the best arrangement for convenience in use. The dog-leg stair has too acute a turn, but it saves space.

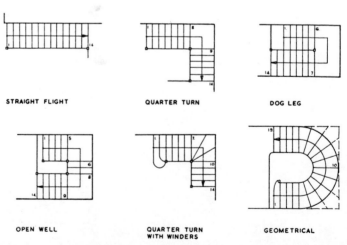

Diagram 139   Staircases: typical plans

Design points

There are many conditions to be fulfilled in a well-designed and constructed staircase. The stair must above all be safe and easy to use. This means that the proportions of treads and risers must be consistent with the theory that it takes roughly twice as much effort to climb up a vertical ladder as it does to walk along the level. The length of an 'average' pace can vary between 550 and 700 mm, and thus the stairs should be in accordance with the formula:

2 x vertical distance travelled + horizontal distance travelled must be between 550 and 700 mm at each step upward, i.e.

$$2 \times \text{rise} + \text{going} = 550 \text{ to } 700.$$

The horizontal distance travelled is the 'going' and the vertical dimension of each step is the 'rise'. The going should be at least 225 mm. If it is reduced below this, the stairs are difficult to descend since there is not enough room for a person's heel to be safely placed in the centre of the tread. A satisfactory proportion for a house staircase would be a 250 mm going with a 180 mm rise, and for a public building a 300 mm going and 140 mm rise would be adequate, to give a slightly less steep flight. The treads and risers must always be of the same proportion in any one flight. Other design criteria for staircases will be found in Part H of the Third Amendment to the Building Regulations.

Stairs should always be well lit, and so solid balustrading presents a slight problem in this respect. Particular attention should be given to lighting at the top of the stairs and to the landings. Electric lights should have two-way control.

Stairs must be of adequate width, i.e. 900 mm between strings. This is based on the assumption that the walking line will be 450 mm from the handrail.

Landings should not be narrower than the width of the stairs, and a handrail should always be provided. Often the top rail of the balustrade is sufficient if it is at the correct height and is continuous. Stairs more than 1 m wide should have a handrail at each side and it should never be necessary to stoop to avoid a bulkhead on the stairs. Diagram 140 illustrates the minimum criteria for a domestic stairs of maximum pitch 42°.

Steps with reducing treads, known in various parts of the country as winders, fliers, dancers or tapered treads should be avoided whenever possible; but where they are inevitable due to the necessity to save space, they should be at the bottom of the flight. All treads should have

154 · BUILDING TECHNIQUES

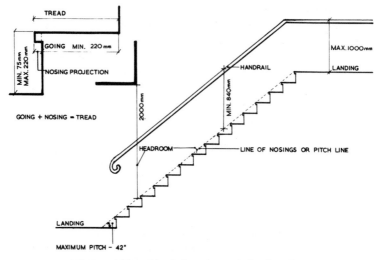

Diagram 140   Criteria in staircase design for a house

a non-slip surface, and the Building Regulations will not allow the tread of a tapering step to reduce below 75 mm at any point for a domestic stairway. Thus the quarter-turn staircase with tapered steps shown in diagram 139 is not to be recommended. The construction, number and position of stairs in public buildings and flats must conform to the statutory requirements for means of escape in case of fire. The regulations determine the widths of the staircases in relation to the number of people using each floor of the building, and all staircases must be of a non-combustible construction. In domestic work, however, the staircase is not required to be incombustible.

Wood staircases

If staircases are to be formed in softwood, which is the least expensive form of construction, the wood should be very carefully chosen to be well seasoned and of good quality. Where hardwood can be afforded, oak mahogany and teak are the most common suitable timbers. In any case the structural members underneath, such as the carriages, landing bearers, etc., can be in softwood.

Wood staircases are usually constructed in flights in the joinery shop, where the treads and risers are fitted, wedged and glued together into the strings, which are the long inclined members each side of the flight.

The simplest way of framing up the staircase is to have newels, which are substantial timbers arranged vertically at the turns of the staircase into which the ends of the strings can be fixed and which can also support the landing bearers. These newels in most cases go down to the floor below and so give direct support to most of the staircase. The string that comes against a wall — wall string — can be plugged to the wall so that one side of the staircase is directly supported. The other string is called the outer string.

Diagram 141 shows a view of the underside of the upper part of a flight. The treads and risers are housed into the strings and wedged. The riser is also tongued into the underside of the tread as shown in diagram 146. Triangular blocks glued on to the underside also help to stiffen the treads and risers. Strings are normally 38 mm thick to allow for the housing of the treads and should be sufficiently wide to cover the nosing, i.e. the projecting part of the tread, and to come down in some cases low enough to mask the supporting member, the carriage, which is fixed afterwards underneath the flight to give support to the middle of the treads. Carriages are not always essential for staircases of 900 mm width, but where they are much wider than this they are necessary. Rough brackets are cut to fit under the treads and are nailed to each side of the carriage.

Diagram 141 also shows tenons formed at the top end of the outer string to go into mortices cut in the newel. The wall string is not secured to a newel, so it does not need such a joint. It will be plugged direct to the wall, and would only require to be shaped so that the skirting round the landing can fit to it.

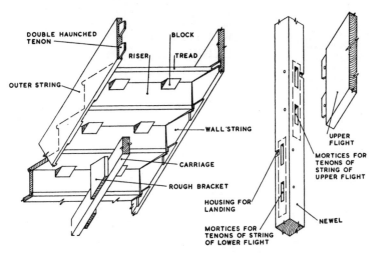

Diagram 141   Staircase details: straight flight and newel

156 · BUILDING TECHNIQUES

The framing for the landing is formed in position when the newel is fixed. The lower landing shown in diagrams 142–144 has a 100 x 75 mm pitching piece screwed to the newel and built into the wall. This supports 100 x 50 mm bearers which run across into the other wall and take the landing boarding direct. These bearers have to be arranged so that they provide fixing for the carriages. The upper landing shown is constructed so as to support the newel, as the newel would be very long if it went down to the floor below. This can be

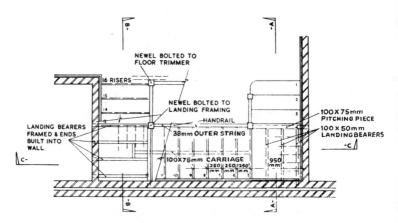

Diagram 142   Plan of open-well staircase

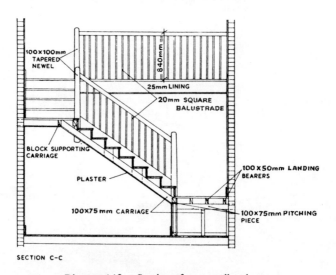

Diagram 143   Section of open-well staircase

STAIRCASES · 157

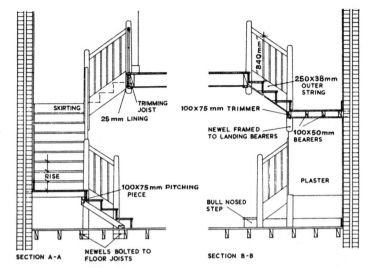

Diagram 144 Section of open-well staircase

Diagram 145 Section of open-well staircase

done by framing the landing bearers with mortice and tenon joints and building the ends into the wall as shown so that it forms a cantilever slab to which the newel can be bolted. The newel at the top of the staircase can be bolted to the floor trimming joist.

The use of newels also simplifies the handrail construction, as these can then be arranged in straight pieces framed between one newel and another. Handrails should be approximately 900 mm vertically above the

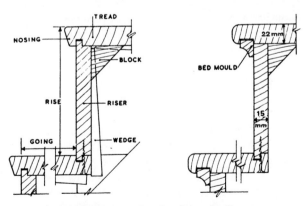

Diagram 146 Tread and riser details

158 · BUILDING TECHNIQUES

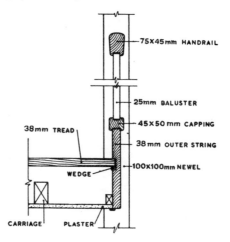

Diagram 147   Section of handrail and outer string

line of nosing and should be of a convenient section to grip. A simple section is shown in diagram 147, which also shows 25 mm balusters. The latter cannot be fixed direct to the top of the string, as it is only 38 mm wide, so a capping of greater width has to be fixed to take both. The space below the handrail can be treated in other ways than with square balusters as shown. A balustrade can be formed of light framing covered with plywood.

In geometrical staircases the strings have to be curved, and this is usually done by gluing laminations of wood together to the shape required. The string has to be very well and accurately made, as it is the chief support of the staircase and does not have the support of constructional newels as in other forms. The handrail is also continuous and shaped to the curve and supported entirely by the balustrade from the string. The construction of these staircases in the past was very ingenious and the staircase hands became very skilled in constructing delicate yet strong geometrical staircases.

Stone staircases

The layouts referred to at the beginning of this chapter are also suitable for staircases constructed of stone or concrete. The simplest staircase of stone consists of steps of square section with their ends built into the walls. They can then be arranged to overlap and so form a convenient series of steps. For the best stone stairs the separate stones are shaped as shown in diagram 148 to have a rounded nosing at the front and to

STAIRCASES · 159

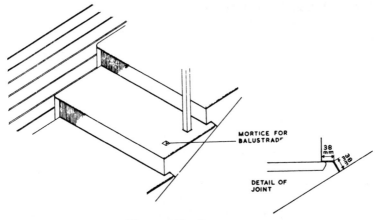

Diagram 148   Stone steps

have a connection to the stones above and below formed with an inclined surface to take and give a direct thrust down the flight. The underside of the stone is often cut away to form a smooth inclined soffit, but in the best work the part of the stone which is built into the wall is left square to give a firm bearing, as shown in diagram 149.

If one side of the staircase cannot be supported, it is possible to rely only on building in one end and on the weight being taken down from one step to another to the bottom. In that case the nosing should be returned round the end as shown in diagram 148.

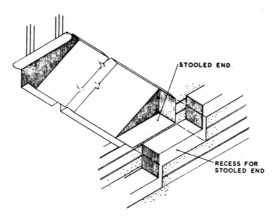

Diagram 149   Spandrel stone step (view from below showing stooled end built into wall)

160 · BUILDING TECHNIQUES

Metal balustrades with metal or wood handrails are commonly used and the vertical members of the balustrade which support the handrail can be let into mortices in the steps and caulked with cement or molten lead. Sometimes metal newels are introduced into the balustrade to support the quarter-turn of the handrail.

Concrete is commonly used as a material for staircases in large buildings. It is most convenient to form the flights as reinforced-concrete slabs bearing between the beams which support the floors and beams fixed to support the landings, as shown in diagram 150. It is then possible to form the steps as cores of concrete on top of the slab. Such a flight can be finished with terrazzo made of marble chippings and coloured cement polished to give a smooth surface, or many other types of finish are possible, such as clay tiles, cork tiles, rubber, etc.

Where flights are very long, steel or concrete beams can be fixed at a rake to support the sides of the flights. In every case the layout of the staircase has to be studied carefully and the arrangement of beams and slabs with proper bearing on the steel or concrete framework or on walls has to be worked out to give support to the steps and landings.

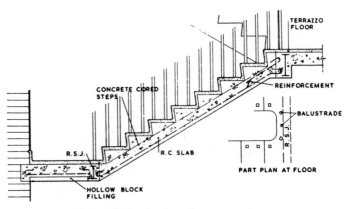

Diagram 150   Section of concrete staircase

Open-tread staircases
Staircases which omit the riser, as in a step ladder, are now quite usual. But if this type of stair is adopted, special care must be taken to ensure that the treads are of sufficient thickness to avoid undue deflection and are of adequate strength to carry the load, as omitting the risers greatly reduces the stiffness of the staircase. Diagram 151 shows a hardwood-

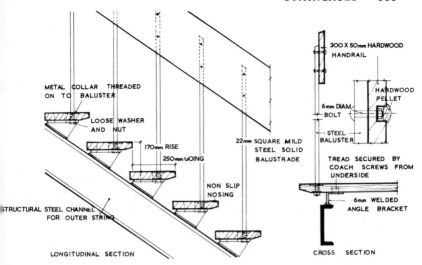

Diagram 151 Open-tread staircase

tread staicase supported on mild-steel brackets welded to channel bearers. The balustrades are constructed from 22 mm square mild-steel rod with bolted fixing through the tread. The handrail must be returned and secured at the head and bottom of the staircase to give stiffness to the balustrading. The Building Regulations specify that in domestic and institutional buildings the gap between the open treads is restricted in an attempt to reduce the risk of accidents.

CHAPTER TEN

Partitions

Partitions

A non-load-bearing partition is a thin lightweight dividing wall which is self-supporting but which must not be used to carry any superimposed loads from the structure. Simple timber stud partitions and plasterboard 'egg-box' construction are two examples. A load-bearing partition may also be comparatively lightweight, and will carry small loads from floor joists or ceiling joists. The most common materials used for a load-bearing partition are lightweight concrete blocks. Alternatively clay blocks to B.S. 3921 may be used. Where the partition is load-bearing it must always be taken down to a proper foundation.

A clay-block partition is shown in diagram 152. Such partitions are much stronger if bonded where they join other walls and if well secured at the top. Such bonding is shown in the diagram. If, however, there is a chance of subsidence, as may happen where a partition is built on a wood floor, it is better to fix it in a chase, as shown in diagram 153. The chase can be lined with paper to prevent adhesion; when the partition settles, it may draw out of the chase slightly and the plaster may crack in the angle, but this is not so unsightly as the cracks that may appear across the face of the partition if it is bonded into the main wall.

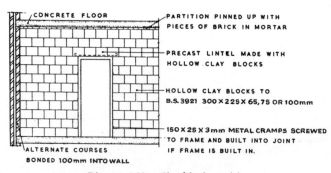

Diagram 152   Clay-block partition

Lintels can be formed over doors by threading rods through the hollows in blocks and filling with fine concrete. Door posts will stiffen the partition if strong metal cramps are built into the joints as shown in the diagram.

Lightweight concrete blocks of the appropriate classification to B.S. 2028 are universally used for partitions. It is advisable, particularly if the storey is high, to carry the door posts up to the ceiling and fix them to the floor above by wedging or by nailing to the side of the joists (diagram 154). In this case the upper part of the door post should be no thicker than the blocks, so that the plaster can be carried over the door post. Strips of expanded metal lathing should be fixed over the post, both as a key for the plaster and to tie in the panel over the door. Alternatively the rectangle of blockwork above the door can be omitted and a glazed frame installed to form a 'borrowed light'. If the width of the door post is equal to the thickness of the partition and its two coats of plaster, a recess can be formed at the back of the frame as shown in diagram 153, which gives greater strength but it is rather wasteful of wood.

To minimize the effect of movement in long stretches of concrete blockwork, construction joints are recommended at intervals of 6 m. These joints are formed by unbonded blockwork with straight butt joints. The blockwork can be 'tied through' at the construction joint for stability by means of thin metal strips say 40 mm wide x 200 mm long x 1·5 mm thick built in at each alternate course which strengthens the break while still allowing movement. Timber cover strips or plaster stops should be incorporated at each construction joint.

Wood-wool slabs are made of long tangled wood shavings coated with cement and compressed. They are very light, have a good key for plaster, and can be cut easily with a saw and nailed. A satisfactory

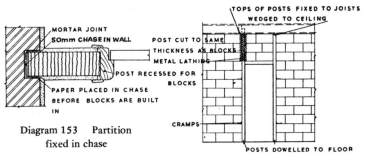

Diagram 153   Partition fixed in chase

Diagram 154   Concrete-block partition

164 · BUILDING TECHNIQUES

method of constructing a partition of wood-wool slabs is shown in diagram 155.

Glass blocks can be built up to form partitions. They are now no longer made in Britain but imported metric-sized blocks are available. A typical glass block is shown in diagram 156 and is built with 6 mm mortar joints, 1 : 1 : 4, cement lime sand mix. The maximum sizes for a

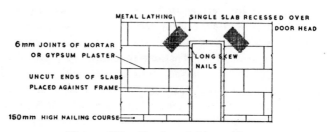

Diagram 155   Wood-wool slab partition

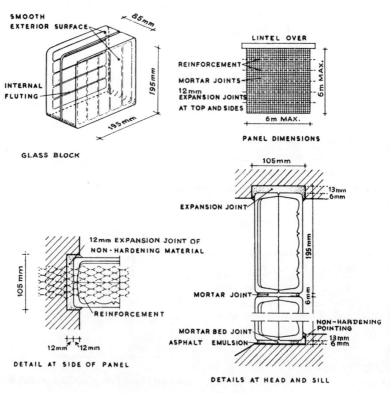

Diagram 156   Glass-brick panel

panel are also shown, it is important that a 12 mm expansion joint should be provided at the top and at the sides. It is nevertheless important to fix the partition or panel at the sides and top, so they should be built in a recess as shown or at least in a rebate. Reinforcement should be used in the joints.

Timber partitions are well understood in traditional carpentry. A simple timber partition is shown in diagram 157, where 100 x 50 mm vertical members are placed at up to 450 mm centres and stand on a sill on the boarded floor. They carry a 100 x 50 mm head which can take light ceiling joists, or can be fixed up to the underside of a floor, which should not, of course, bear directly on the partition. The spacing of the studs depends upon the covering material, and for such a partition plasterboard would be the normal choice. The sheets are 2·4 x 1·2 m so that studs at 400 mm centres would be satisfactory.

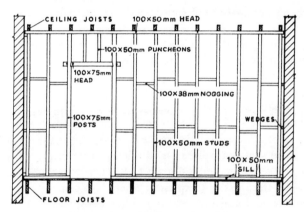

Diagram 157 Timber-stud partition

To stiffen the partition, horizontal pieces 100 mm wide should be cut to fit tight between the studs and can be arranged as shown in the diagram. It is important that wedges or blocks be placed between the last stud and the wall to complete the stiffening. If sheet material is used, these horizontal members also should be arranged in line (not staggered as shown) and at intervals to suit the sheets. The members that trim round door openings should be stronger, such as 100 x 75 mm and should be well framed together.

'Eggbox' partition

Plasterboard can also be obtained as a prefabricated 'dry' partition which consists of thin cardboard in the form of a square cellular core of 'eggbox' construction faced on each side by plasterboard to form a rigid readymade unit 38 mm, 50 mm, or 57 mm thick in wall-panel sizes. The plasterboard can be 'ivory surfaced' ready for decoration or 'grey surfaced' to receive a plaster skim. Each section of the partition is jointed over a continuous timber batten which also occurs at junctions and angles. Additional timber facing plugs can be tapped into the core of the partition where required as fixings for skirtings and linings. The 57 mm units can be used to form a double partition with a centre cavity. This type of construction is useful where increased sound insulation is required since the cavity can be filled. Diagram 158 shows details of both types of partition. The term 'dry' construction is only really applicable when the construction is ready for decoration without the application of a wet plaster skim coat.

Dry lining
The technique of applying plasterboard sheets direct to the wall, known as 'dry lining' is shown on diagram 158. A level surface is first obtained by fixing impregnated fibreboard 'dots' on the uneven brick surface. The plasterboard sheet is then fixed on plaster 'dabs' set in the spaces between the 'dots'.

Internal finishings
Skirtings are necessary to prevent damage to the plaster or wall covering near the floor. Where hard plaster is used it may be sufficient to form a painted band, but normally a wood skirting is formed fixed to rough grounds plugged to the wall (diagram 159). These rough grounds should be the same thickness as the plaster. Where high skirtings are required they have to be formed of several members, as shown on the right of the diagram.

Metal skirtings can also be used for buildings. A typical section is shown.

Walls can be lined with sheets of plywood, wallboard or plasterboard. There are several advantages over ordinary plaster, such as better insulation and the fact that a room so lined will heat up more quickly. The use of some of these sheet materials is restricted to certain areas by the fire requirements of the Building Regulations and they must also be acceptable under the surface spread of flame classification. Trade literature will give the technical information regarding performance which can be compared with the performance requirements of the Building Regulations to enable a satisfactory choice to be made. Most sheets have to be fixed on battens to secure a true face for the fixing

PARTITIONS · 167

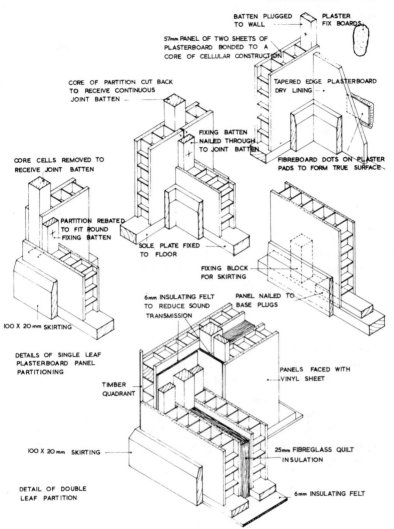

Diagram 158   Plasterboard panel partitioning

and to keep them clear of the wall. Battens should be 50 mm wide and are plugged to the wall behind. They should be arranged under each edge of every sheet. Additional intermediate battens are required to support the sheets in the centre, the placing varying according to the type of sheet. Plywood should be supported at 300 or 375 mm intervals.

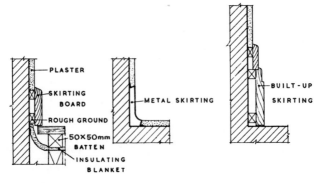

Diagram 159   Skirtings

One problem is the finish of the joints between sheets. The simplest and most satisfactory method is to fix a wood fillet over the joints, in which case any slight variation in the gap between the sheets does not matter. Such a cover fillet is shown in diagram 160. Too many cover fillets, however, are unsightly and a pleasanter appearance is obtained with a V-joint which can be formed in wall-board or plywood sheets. The sheets have to be fitted very carefully, however. Many patent metal cover strips are available, a typical one being shown in the diagram. This is fixed on one side of the tee only, the fixing being covered up by the insertion of the next sheet. In this case no fixing through the sheets is necessary, but in the case of a cover strip the sheet would be fixed every 100 or 125 mm along the edge with a wire nail. In the case of the V-joints the sheets should be fixed with panel pins, the heads being punched in and stopped with coloured stopping. It is very important not to bruise the sheet with the hammer.

As will be seen from the elevation in diagram 160, the covering of a wall has to be carefully planned so that the cutting of sheets to waste is kept to the minimum, but at the same time appearance must be considered. In some cases the positions of windows and doors should be planned beforehand to allow for this economy in the cutting of sheets.

Plastering
For internal plasterwork two or three coats may be used, the third coat only being necessary on very irregular surfaces, such as on the inside of old brick walls, stone walls or on buildings being altered for conversion. In three-coat work the first or rendering coat is usually cement and sand, 1 : 3, and should grip the backing securely and fill up the hollow

PARTITIONS · 169

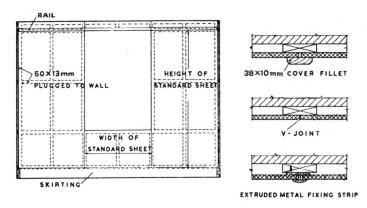

Diagram 160   Wallboard wall lining

places. The second or floating coat may be cement and sand or gypsum plaster and sand, and this coat can be brought up to a true surface so that the third or setting coat, which can be plaster only, need only be thin. Lime plaster was much used in the past for the setting coat and also in the other coats. The lime sets chiefly by drying, and air, on reaching the particles of lime, turns it into calcium carbonate, but this is a long process. For this reason gypsum plasters, which set by action with the water with which they are mixed, have largely replaced lime in plastering. These plasters, however, produce a very hard, cold finish, which is not always desirable, and the use of a proportion of lime in the setting coat and even in the other two coats can be specified. Great care has to be used, however, in adding lime to certain types of plaster, because of the likelihood of unexpected chemical reaction.

Three-coat work on brick walls should be a total thickness of 15 to 20 mm and on partitions it may be a little less. Two-coat work, which is possible on certain partitions, would be 10 to 13 mm thick.

Pre-mixed plaster
Pre-mixed plaster is now widely used by Contractors since it guarantees that the materials can be precisely in accordance with the Specification throughout the Contract. It simplifies ordering and eliminates the mixing of raw materials. A lightweight exfoliated aggregate is used which produces a finished plaster which provides a very high fire resistance with much-improved insulation characteristics. This type of plaster is particularly useful in two-coat work on concrete surfaces.

170 · BUILDING TECHNIQUES

Ceilings

Ceilings used to be formed with wood laths 20 x 6 mm spaced 12 mm apart and nailed up to the underside of the floor joists or to separate ceiling joists as shown in diagram 161. The lathing supported the plaster, which would be about 15 mm thick and consist of three coats, the first a rendering of lime and sand to which horse hair had been added, the second a floating coat mostly of lime and a little sand, the third a thin setting coat of lime only. This ceiling is heavy and takes a long time to dry out, though the addition of cement hastens the set, but this method has now given place to metal lathing or plasterboard ceilings, although it will still be found in buildings which are being altered or converted.

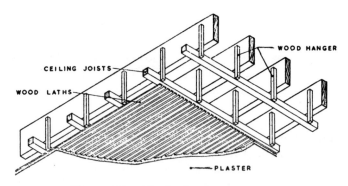

Diagram 161   Wood lath and plaster ceiling

Metal lathing, which is like heavy, close wire netting, is supplied in rolls and fixed to the joists with staples. Two- or three-coat plastering as described above is then carried out, but with more cement to give a quicker set.

Plasterboard 9·5 mm, 12·7 mm or 19 mm thick, consisting of a sandwich of gypsum plaster between two sheets of cardboard, is supplied in sheets as follows: Wallboard, 600, 900 and 1200 mm wide and 9·5 and 12·7 mm thick; Plank, 600 mm wide and 19 mm thick; Baseboard, 914 mm wide and 9·5 mm thick; and Lath, 406 mm wide and 9·5 or 12·7 mm thick. The sheets are available in lengths in a range of sizes between 1200 and 4800 mm according to type; it can be cut and fitted and nailed up to the joists. This gives a smooth, true surface which can be covered with a coat of special plaster as little as 5 mm thick. The joints between the boards must be surfaced with a thin strip

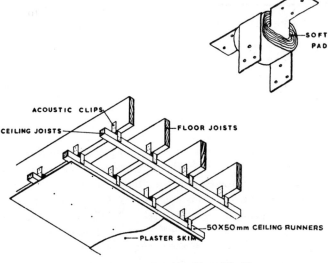

Diagram 162   Plasterboard ceiling with skim coat

of open weave hessian soaked in wet plaster before the 'skim' coat is put on, otherwise cracks will appear along the lines of the joints. The hessian 'bandage' is known as 'scrim'. The general technique is shown in diagram 162, which also shows acoustic clips used for hanging the ceiling joists from the joists of the floor above. They are made of two strips of metal with a soft pad between so that sound vibrations are not so easily transmitted across the connection.

CHAPTER ELEVEN

Applied Finishes

Paints
Paint is composed of solid matter, known as pigment, which is held in suspension in a liquid medium. When spread over wood, metal, plaster, brick, concrete and other surfaces it dries to a solid film. Thinners, driers, extenders and other ingredients may be included to vary the properties of the paint during application and drying.

The function of most pigments is to give colour and to form a covering over the surface. Some pigments have, in addition, rust-inhibiting properties and others offer resistance to chemical attack. The medium binds the pigment particles together, enabling the paint to be spread in liquid form and largely determines its type and general characteristics.

Reasons for painting
Painting preserves and protects the materials used in building. Wood and metal in particular would quickly deteriorate without the protection of paint. The natural elements, rain, frost and sun cause the most damage but in an industrial atmosphere chemical fumes are also very destructive. It is especially necessary to protect the light metal or timber framework used in many modern industrial and system buildings.

Paint, of course, makes a valuable contribution to the appearance of the building. Wall colour is principally a matter of aesthetics but colour is also used to identify hazards and danger points.

It must be understood that paint protects only so long as the film remains intact. Once the surface has been broken, the way is clear for the deterioration of the protected material.

On timber, the sheen and texture of the paint is affected by the type of wood and with clear finishes the natural colour and grain of the wood is intensified. The surface coating also reduces the absorption of moisture vapour which consequently limits the tendency of the timber to warp or split. Thus if the paint completely covers and seals the timber, dimensional change is prevented and thus the weathering is slowed down.

The pigment used in paint is capable of absorbing ultraviolet light. Clear finishes, which do not as yet include an ultraviolet barrier as good as pigment, will deteriorate much more quickly than pigmented paint because of this should be borne in mind when the use of a clear finish is contemplated for hardwood in an exposed position outside.

Types of paint
The method of classifying paints under the headings of oil-based, water-thinned, cold-cured and clear finishes will have the most practical value, though these distinctions are not always clear-cut.

Oil-based paints
The simplest oil-based paints, often referred to as 'readymixed', consist of a vegetable drying oil, such as linseed oil, with pigment driers and turpentine or white spirit. They are the kind that decorators used to 'knock up' for themselves, applying several coats and following with coats of varnish to provide gloss. Although these simple paints are slow in drying, relatively soft and tend to show brushmarks they are still used on surfaces that cannot be prepared sufficiently well to receive more up-to-date coatings, or where appearance is not of prime importance. Some primers are of this type of oil base.

By combining a natural or synthetic resin with the drying oil, and by heating the oil, oleo-resinous paints are produced. As these dry to a hard glossy film which does not need subsequent varnishing, they are often referred to as hard-gloss or varnish paints.

Alkyd-resin paints also contain drying oil, but this is reacted chemically to produce a resinous vehicle rather than a simple mixture of resin and oil. These paints are a great improvement on the older types. Their method of manufacture can be precisely controlled and is independent of the variables arising from the use of natural resins. When applied over the correct primer, and following the correct procedure, they are suitable for almost every building surface, providing easy flow, quick drying and a tough, lasting film.

Modern high-quality gloss finishes are based on alkyd medium and they can be formulated as flat, matt, eggshell, semi-gloss and full gloss. Some primers and undercoats also have an alkyd-resin base.

Water-thinned paints

A true distemper is of the non-washable variety — whiting bound in size — and cannot even be lightly wiped over with water without the danger of dissolving the size and removing the film.

Water paints, sometimes called oil-emulsified distemper, are simple emulsions of oil and water. Although they are described as being washable, cleaning can be carried out only by wiping, not by vigorous rubbing.

Emulsion paints are based on dispersions of synthetic resin in water. The resin includes styrene butadiene, acrylic and polyvinyl acetate (PVA), singly or in combination. Polyvinyl acetate or PVA acrylic copolymer emulsions are the most commonly used in Britain. Emulsion paints can be formulated to be suitable for either exterior or interior use. Those for interior use resist steamy atmosphere reasonably well, last longer than water paints and can be washed. Some types of emulsion paint dry with a fairly high sheen, while others are sometimes though perhaps not wisely used as primers for wood or as an undercoating for conventional gloss finishes.

Other water-thinned paints are stone and cement paints which are formulated for use on exterior cement rendering, and the heavy-duty emulsions. There are also texture paints for relief work on inside walls and ceilings.

Cold-cured materials

The materials described above dry mostly by absorbing oxygen (oxidation) in the case of oil paints, or by evaporation in the case of water-thinned paints. In recent years a new class of coating has been developed in which the change from liquid to dry state is accomplished by a process of curing or, in chemical terms, 'cross-linking'. Curing may be induced by heat (stoving), but with paints for site use on buildings this is not possible and so curing at room temperature is employed. This 'cold-curing' is induced by adding chemical agents, known as activators or catalysts; the most widely used coatings of this type are based on epoxy and polyurethane resins.

While there are some differences between the properties of the two types, both have exceptional hardness and have an outstanding resistance to solvents and chemicas that would destroy or damage normal paints. Both are supplied in two-pack form, the components being mixed together immediately before use. They then remain usable for several hours, after which they must be discarded. They require meticulous surface preparation and they may not be compatible with old paint on the surface. They are also sensitive to temperature and humidity during application and curing. For these reasons their use is generally confined to situations where normal coatings would be inadequate, such as where there is likely to be severe chemical attack. It is advisable, before using coatings of this type, to seek guidance from the paint manufacturer and to follow his instructions implicitly.

Epoxy-resin and polyurethane paints are also obtainable as one-pack coatings but these do not possess quite the abrasion and chemical resistance of the two-pack types and, apart from greater speed of drying and slightly greater chemical resistance, are not markedly different from alkyd-resin based paints.

Clear finishes

The practice of applying varnish over paint coatings to produce a glossiness is rarely followed nowadays, though the increasing use of hardwood joinery and timber cladding has resulted in a demand for clear protective finishes.

Clear alkyd, oleo-resinous, polyurethane and epoxy varnishes and lacquers may all be employed for this purpose. For exteriors, the alkyd and oleo-resinous types are still superior to the more sophisticated polyurethane and epoxy-resin in that the latter are sensitive to surface preparation and application conditions; they may also prove troublesome when the time comes for renewal of the coating and then the new coating may not adhere satisfactorily to the old. On floors and bench-tops, subject to exceptionally hard, abrasive wear, polyurethane varnishes are excellent.

For interior work where clear glossy or satin finishes may be required on timber, there are quick-drying lacquers, often on an alkyd-urea formaldehyde base, which give a finish similar to that of french polish, but they are more resistant to abrasion and alcohol spillage. This type of finish is simpler to apply than french polish and eliminates the need for specialist labour.

To achieve reasonable durability on exteriors three or four coats of varnish are necessary. As this is expensive a demand has arisen for cheaper systems. An alternative is a simple mixture of linseed oil, paraffin wax, pigment (if required) and a fungicide, a mixture that was originally called Madison sealer, but which is now available under various proprietary names. One or two coats of this type of seal will give good protection for two or three years. Its main disadvantage is that it does not dry hard and tends to pick up dirt.

Thixotropic paints
These have the consistency of a thick jelly. The action of brushing temporarily destroys the thixotropy and enables them to be spread, after which they re-gel until dry. This means that there is less likelihood of runs and sags developing and they do not readily drip or spill. They are sometimes known as one-coat paints because a film of approximately one and a half times the thickness of a normal coat of paint can be achieved in one application. They are not popular for commercial use.

Bituminous paints
These are based on bitumen, tar or pitch and, provided that their bitumen content is reasonably high, they have good water and chemical resistance. In general, however, their colours are drab and uninteresting. They tend to crack and lose gloss under a hot sun and cannot subsequently be recoated with an oil paint unless a sealer is applied first. Even then, the paint coating will be liable to craze owing to he relative softness of the underlying bitumen.

Chlorinated rubber paints
These need special primers and undercoats and are used where a high resistance to water, inorganic (not fatty) acids and alkalis is demanded. They are soluble only in their own solvent and so subsequent recoating often leads to 'lifting'. Specialized techniques of application are necessary because of their tendency to bubble and form pinholes.

Multicoloured paints
These are composed of different coloured particles which are prevented from merging by being enveloped in a special cellulose medium. When sprayed on to a surface under pressure (they cannot be applied by brush) they produce, in one application, a multicoloured finish usually consisting of two or more spot colours. The finish is tough, hard-wearing and easily cleaned and so it is often used on the interior walls of public buildings. It is not suitable for exteriors.

Other paints for special purposes
Floor paints which have a special medium with hard-wearing qualities, heat-resistant paints, again a special medium, and anti-climb paint with a non-drying medium.

Preservative coverings

Apart from film-forming paints and varnishes there are protective materials composed of fungus-destroying chemicals which penetrate the pores and cellular structure of timber. Creosote is a simple form of this type of covering. It is cheap and quickly applied; but its disadvantages are that it is obtainable only in brown and it tends to leech out leaving a bleached surface when the timber is exposed to weather. If an oil paint is subsequently applied the solvents in the paint will activate the old creosote and cause 'bleeding', unless the surface is sealed first. Proprietary wood preservatives will penetrate deeper than creosote. They will last considerably longer and are obtainable in a limited but pleasant range of colours. Some of them will take oil paint without the use of a sealer.

Preparation of surfaces

The more intimate the contact between paint and substrate, the better the paint adhesion will be. If the surface is dusty or powdery, not only will the paint fail to stick properly but its appearance will be impaired. If the surface is greasy, the paint may be slow in drying and in extreme cases will not dry at all.

It is essential, therefore, for a surface to be thoroughly cleaned before painting and then for all traces of cleanser to be removed with clear water. Adequate time must be allowed for drying, especially where there are open joints, angles or crevices which will hold water for a long time.

As timber on a site is often inadequately protected it will rapidly absorb moisture, and this is particularly so with kiln-dried timber. When dry, it should therefore be primed as soon as possible. Joinery is better primed before it leaves the joiner's shop. The primer must be carefully applied to the edges and endgrains which may be inaccessible after fixing.

Iron and steel must be cleaned thoroughly before painting, and priming should then be carried out immediately. The cleaning is usually carried out by mechanical means using 'needle guns' or blast-cleaning. New zinc and new galvanized iron require treatment with a mordant solution before priming, or else they require to be primed with calcium plumbate primer. After the surface has weathered, zinc chromate primer may be used as an alternative, provided that the surface is clean.

Copper and lead provide a poor foundation for paint and, where decoration is required, finishing coats only (no primer or undercoat) of an alkyd-resin paint should be applied after the surface has been rubbed down and wiped over with a grease solvent. After shop-primed radiators have been fitted, damaged parts must be spot-primed with zinc chromate primer and finishing coats of paint applied on top. Undercoats are not recommended.

Plaster, brick and concrete surfaces are inclined to be alkaline. When new, or in the presence of moisture, salts may be brought forward which will attack the medium in an oil paint. If it is impossible to postpone painting until drying out is complete, paints based on epoxy resin that will interfere little with the drying, such as distemper or emulsion, should be used, or those that are not readily attacked by alkalis. The use of alkali-resisting primer under an oil paint will only minimize the risk of trouble, not completely eliminate it.

Backs and edges of fibre building boards should be painted before fixing to prevent warping or buckling. It is better not to rub down the faces because abrasion will score the surface. Asbestos cement and asbestos/wood sheets are alkaline by nature and require similar treatment to that indicated for new plaster. Flat or matt paints are best for acoustic boards which are made of a fairly soft fibre.

Many fibreboards are sold already primed by the makers, necessitating only the touching in of damaged parts and cut edges.

Old work that has previously been painted does not need a primer, except for the touching in of parts that have blistered or flaked, and these should first be scraped clean and the stepped edges feathered smooth with abrasive paper.

The paint system

The primer is the foundation of the whole paint system, and so the paint manufacturer's instructions as to the use of the correct one should be followed faithfully. Some paints and some surfaces require a special primer, others only a thinned or matt coat.

Gloss finishes require an undercoat over the appropriate primer (except on lead and copper) and two finishing coats, or alternatively two undercoats and one gloss — to build up a film thickness necessary for minimum protection outside. On interiors one finishing coat may be omitted.

Eggshell and flat finishes are suitable only for interior use. The undercoat may be omitted and the finishing coat applied directly over the primer to save cost.

Where emulsion paints, water paints and distempers are applied over plaster, brick or concrete, a first well-thinned coat is needed to equalize porosity, and this should be followed by a normal coat thinned as directed on the container label. With size-bound distempers, care needs to be taken to ensure that the second coat does not activate and 'lift' the first coat.

The system of application of clear finishes, polyurethane, epoxy resin and chlorinated rubber paints varies widely. The instructions of the manufacturer should therefore be followed in each case.

The same amount of labour is required to apply a cheap paint of doubtful quality as for a good paint made by a reliable manufacturer, and as the cost of labour involved is about four-fifths of the total decorating bill, it is economic to use only paints of repute.

Recommended coatings for alkyd-resin paints on outside work are one primer, two undercoats and one finishing coat, or one undercoat and two finishing coats, and this will give a film thickness of about five-thousandths of an inch. Subsequent painting can be confined to undercoat and finishing coats.

A good alkyd-resin paint correctly applied to a well-prepared exterior surface facing South, where it will be subjected to the most sun, should last 4 years, and an emulsion paint about the same time. On surfaces facing North, its life will be half as long again.

An alkyd-resin paint that has not blistered or flaked, through a fault in the substrate, will finally wear by 'chalking'. This is the result of the surface pigment being denuded of its binder but still remaining on the surface in the form of a bloom. When this bloom is washed off, the surface will be ready for repainting, and the most economic time to repaint is before chalking has reached such a stage that the undercoat is exposed.

The period between interior repaints is naturally governed by the type of work carried out within the building.

Paint application

The main tools of a painter are brushes, rollers and spray equipment.

More paint is applied by brush in the decoration of buildings than by any other method because a brush will reach into corners and intricacies inaccessible to other tools. Best-quality paint and varnish brushes are made of pure hog bristle which has a natural curve, taper and flag, to hold the paint well and to enable a high standard of finish to be achieved. Cheaper brushes include short bristles or a mixture of hog bristles and other filling materials.

Tapered nylon filament used as brush filling has the advantage of harder-wearing properties and it is cheaper than hog bristle in the larger sizes of brush. Disadvantages compared with hog bristle are its inferior paint-holding qualities and its inability to wear to a satisfactory shape. Nylon brushes are admirably suited for use on rough surfaces.

The use of rollers for paint application saves a great deal of time. They are made in sizes from about 50 mm to 350 mm wide and apart from the usual cylindrical type they can be designed in different shapes, convex, concave or corrugated – to suit unusual contours. Those covered in mohair serve all purposes. Lambswool rollers hold more paint but as they are apt to shed fluffy bits when used with a viscous material, they are suitable only for water-thinned paints. Foam-covered rollers give the smoothest finish, but can only be used on surfaces which are already comparatively smooth.

Owing to the time taken to mask out parts that do not require painting, spraying is economic only where large areas are involved. So-called 'airless' spraying employs compressed air to drive a reciprocal ram pump, which lifts the paint from a container under pressure. The paint under pressure is passed through a metal-reinforced hose to the gun, the control of which consists of a needle or ball-type cutoff valve, allowing the paint to pass out when the trigger is squeezed. When correctly used, airless spraying does not result in too much overspray. It is fast, covering up to $10 \, m^2$ a minute, and the spray will take high-solids materials, such as rust-preventative compounds. Another, more conventional spray gun with separate air supply from a power compressor is illustrated in diagram 163. Electrostatic spraying is still in

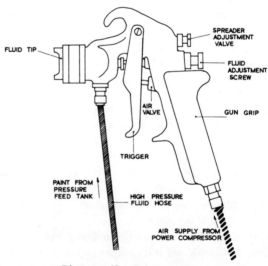

Diagram 163   Paint spraygun

an experimental stage for the application of paint to buildings. It relies on the principle of charged paint particles taking the shortest path to an opposite charge (an earthed object). Paints used in electrostatic spraying must have the right electrical characteristics, and as the special types of paint used do not operate satisfactorily in handguns, this technique is thus employed largely in industrial painting.

Painting defects
When anything goes wrong with a paint system it is very natural for a painter to blame the paint, though it is, in fact, rare for a reputable paint to be at fault. The most common painting defects are listed at the end of the chapter on pages 183–184.

Wallpaper
The standard roll of wallpaper contains approx. 10 m of material 520 mm (52 cm) wide when trimmed.

Surfaces to receive wallpaper should be very carefully prepared even though the covering, particularly if textured, will hide minor surface defects. The aim of the preparation is to produce a dry surface free from cracks and nibs and having a slight suction. Thus newly plastered walls, damp walls, or walls where efflorescent salts appear, require special preparation, and in certain cases it is advisable to line the walls with a lining paper to provide a good foundation, particularly where it is intended to use a heavyweight or handprinted finishing paper. The lining paper should be hung with butt joints across the wall, with the joints parallel to the floor. On old plaster a coat of size ought to be applied, this will prevent unduly absorbent patches by equalizing the suction and so assist in adhesion. The sized surface also makes paperhanging easier by allowing a certain amount of 'slip'. Lining papers can be inexpensive white or off-white pulp paper, or strong waterproofed brown linings, or even metal foil or laminated lead as a temporary measure on damp walls. If the wall is badly cracked or where it is necessary to cover up joints as in wallboard linings, then a calico-faced lining can be used. The correct type of paste should be used and the maker's advice should be taken. Use of the wrong adhesive will result in discoloration and mould growth on the surface. It is desirable for hygienic reasons that old wallpaper be removed before the surface is repapered.

Also available are special types of wallpaper such as washable and readypasted. Woven fabrics such as hessian or fibreglass material are also used for wall coverings.

Plastic wall coverings

Vinyl wall coverings are now much used for high-class work. The covering consists of a PVC film either patterned or texture embossed and laminated to a paper backing. The surface texture and appearance is similar to a good-quality wallpaper. The walls to be covered must be clean and dry and prepared as described for wallpaper. Since vinyl wall covering is a waterproof material, special adhesives must be used which contain a mould inhibitor.

Standard colours

Most manufacturers produce colours conforming to those shown in B.S. 4800. This standard range of colours was devised to provide colours which could be satisfactorily used together in schemes of decoration. The colours are arranged in categories of lightness, hue and strength. The colours are coded within these categories for easy comparison and each colour is numbered accordingly. Thus the colour is specified by a number avoiding the confusion of the subjective inference of a name. For selling to the general public, however, the manufacturers still retain colour names since a well-chosen emotive name is an important sales consideration. In addition to the B.S. colour reference, a colour in the standard range will also be given an approximate Munsell reference. The Munsell system is internationally accepted as a means of colour identification and clarification for many purposes apart from decoration. The classification of the colour of subsoils is one example.

The Munsell system assesses colour under the three headings of hue, value and chroma. Hue indicates the pigmentation, distinguishing say red from blue, value is a measure of the lightness of a colour, and chroma shows the strength of the colour. This will be seen to be a similar method in broad outline to the B.S. classification. The difference, however, is that the B.S. system identifies a particular colour while the Munsell merely indicates the position of a colour in an infinite system.

APPLIED FINISHES · 183

Defects in Painting

Defect	Cause	Cure	Prevention
Bittiness	Dirt on the surface or on the painting tools; or bits of skin which have been stirred into the paint instead of being strained out	Rub smooth when the surface is hard and apply a fresh finishing coat	Strain the paint and clean the tools
Bleeding (staining by a soluble and coloured substance underneath the paint finish)	Applying an oil paint over bitumen or creosote, and previous application of red paints containing dyestuff pigments	If removal of the coating back to the substrate is not practicable, apply a sealer before repainting	Remove material likely to cause trouble, or use a sealer
Blistering	Moisture or solvents trapped in the surface	Strip and repaint if the blisters are numerous; scrape and feather the edges if they are few	Allow thorough drying of all surfaces that have been washed or recently painted, and avoid working in inclement weather or under the direct rays of a hot sun
Blooming (mist or haze giving apparent loss of gloss)	Moisture or contaminants in the atmosphere while work is proceeding	Polish with a soft rag or, in extreme cases, wash or scrub. (The bloom may, however, return)	Apply final coats in clean, warm and dry conditions
Cissing (shrinkage of the paint film into craters or gathering into blobs)	Grease or silicone polishes on the surface, or exudation from an over-oily undercoat. Water-thinned paints sometimes ciss when applied to oil-based paints.	Rub down when hard and recoat	Clean the surface thoroughly before painting

Defects in Painting (continued)

Defect	Cause	Cure	Prevention
Curtaining, running, sagging	Uneven application, or painting over a wet edge that has started to set	Rub down when dry and recoat	Brush out paint evenly, especially on moulded or riveted surfaces, and 'pick up' wet edges quickly
Efflorescence (white crystalline salt deposits on new plaster, brick and cement surfaces or on similar older surfaces that are damp)	Alkaline salts which disrupt the paint film and discolour it	Wipe off if an emulsion or distemper film has not been harmed, and strip an oil paint film, wipe and leave to dry before recoating	Allow the substrate to dry out thoroughly and wipe off deposits as they appear
Flaking and peeling	Expansion of a substrate, a powdery or badly chalked old paint or a greasy surface, or the use of an unsuitable primer or undercoat	As for 'blistering'	Ensure that the surface is clean, dry and firm before painting, and use primer and undercoat recommended by the manufacturer
Grinning (poor opacity)	Spreading paint too far, using an undercoat of unsuitable colour, or attempting a wide colour change with too few coats	Apply further finishing coats	Use the recommended undercoat, and more undercoats when there is to be drastic colour change
Saponification (soft, sticky masses, exuding drops of a brown gummy liquid)	Attack by a strong alkali	Strip and recoat	Keep strong alkalis away from oil paints and ensure that surfaces are free from alkali
Shrivelling	Applying paint unevenly or too thickly	Rub down and recoat	Thoroughly brush out a paint coating and do not allow it to build up along points or edges

CHAPTER TWELVE

Building Maintenance

The term Building Maintenance is used here with three separate definitions in mind. First, there is the work carried out by the Contractor on new buildings, six months (or other agreed period) after the building is completed. Secondly, there is the routine work that ought to be carried out during the life of the building by the owner or occupier to prevent deterioration of the fabric, and thirdly, there are the repairs necessary in old buildings to prolong their life.

New work

The Building Contractor under the terms of the Building Contract will agree to put right any defects, shrinkage or other faults which may occur within a limited period after the building is completed due to the materials or workmanship not being in accordance with the Specification. This period is usually six months on new contracts, but may be reduced by agreement to three months for alteration work, and is known as the Defects Liability Period.

As building techniques become more complicated, new problems arise and very often new faults appear. In general, however, movement defects are the most common, and one or two taken at random are given below with comments.

Plaster cracks, due to shrinkage caused by drying out and movement of the backing material, occur most frequently at junctions in the plaster coat, and can be minimized by the correct use of hessian scrim to reinforce the plaster at these points. One of the major delays in traditional building is the time taken in waiting for the moisture used in concrete, mortar, and plaster to dry out, and the shrinkage and cracking is a side effect of this drying out process. For both these reasons 'dry' construction is now used wherever possible and the increasing use of industrialized building will speed the process.

Timber is also likely to twist and shrink on drying out and, in particular, doors which connect rooms of differing humidity, such as between kitchen and hall and outbuildings, will be liable to movement.

All doors should be hung on three hinges (a pair and a half), rather than two, to give an extra fixing point to restrain movement. Floorboards are also subject to drying out causing the joints to open. In order to reduce the movement of timber, the building should be made weathertight by glazing as soon as possible. Timber fittings, such as counters or cupboards, should not be delivered until the building is heated and the temperature and humidity reach the condition to be expected when the building is in use. Timber used in fittings is specified to be of a given moisture content (usually between 9 and 14 per cent) and so it is folly to deliver the units if the building is cold and damp. Sheet materials such as fibreboard and hardboard are highly susceptible to moisture movement and they should be correctly protected and stacked flat with adequate continuous support so that they do not twist out of shape.

Sheet floor coverings, such as linoleum, tend to stretch after being fixed and so they should be 'laid loose' with overlapping joints for about three weeks before finally being trimmed and then fixed.

Clay floor tiles such as are commonly used in laboratories or circulation areas where heavy wear is expected are subject to thermal movement, and this is particularly noticeable where the tiles are laid as a covering to upper floors where the heat from the room below affects the underside of the tiles. Small lateral movements can cause a surprising amount of lifting, so these tiles should be laid on a 'floating' screed at least 55 mm thick, separated from the main construction by a polythene membrane.

Wood blocks are very susceptible to changes in humidity and temperature, and the resultant movement is cumulative along the grain of the timber and can amount to as much as 25 mm in a 3·5 m length. It is particularly troublesome if the blocks are laid 'brick bond' pattern with the grain parallel in each block. It is better to lay the blocks 'basketweave' or herringbone' pattern so that the direction of the grain is reversed every few blocks. In any case the blocks should have an open expansion joint around the edge of the floor, under the skirting, at least 13 mm wide.

Thermoplastic floor tiles give little trouble from movement but water pressure in the sub-soil will cause these to lift and so a waterproof membrane is advisable in every case.

The opening lights of a timber window may distort and will require 'easing'. The distortion is invariably caused by water penetration behind a broken paint film which causes the timber to swell. The correct treatment is to plane off the timber until a correct fit is obtained, plane off a further shaving to allow for the new paint film and then immediately prime the bare woodwork to seal it. Otherwise the trouble will recur.

Built-up felt roofing may fail, due to excessive movement of the substructure, and it is a wise precaution to detail an expansion joint in the roofing at about 10 m intervals. This can take the form of a strip of roofing felt, laid loose over a rubber-hose former, and then stuck down at the edges. Alternatively a copper strip can be bent up in the form of a continuous 'hinge' and laid under the first layer of felt.

Routine maintenance

In addition to the inspections that will be made by technical personnel on heating equipment and other services from time to time, there are many routine checks that ought to be carried out by the building owner to make sure that conditions are not set up which will make repairs inevitable.

Dust and dirt in the atomsphere which are dissolved out by rain, settle in the bottom of gutters and form a thick sediment, and together with leaves blown by the wind and small twigs and grass carried by birds can quite soon cause a blockage in the rainwater pipes. Thus, gutters should be cleaned out annually and the outlets checked to see that they are clear. A check should also be made to see that ventilator outlets or inlets are not blocked by plants or creepers. Cistern overflows and ball taps should be inspected regularly to see that everything is in order.

Older property: maintenance reports

Local Authorities and other building owners arrange for an experienced surveyor to inspect their property at regular intervals. The period of inspection varies from twelve months to five years and the surveyor will prepare a report on the condition of the property which is sometimes referred to as a 'Schedule of Dilapidations'. Intending building owners also often ask for a report on the condition of a property before purchasing. Reports on property should describe the construction and should then detail its condition and note all the defects that are apparent. It is often difficult to assess the full effect of defects such as dry rot or woodworm unless the construction is opened up by removing panelling or floorboarding. In the case of a sale of property, this would not normally be permitted by the vendor, and so the surveyor should set down clearly the limitations of his report. The inspection of the building should be carried out systematically and the headings given below will be a guide to the contents. The first part of the report should deal briefly with the building in general terms as follows:

History: A history of the building together with a list of the accommodation by room names.

Construction: A description of the method of construction of the building with the materials used.

Fittings: A list and description of the built-in fittings.

Services: A description of the services installed, i.e. electrical, central heating and drainage, indicating materials used. A comparison of the method of installation and materials used with present-day practice, will be a guide to the age and condition of the parts of the installation which cannot be examined.

The second part of the report should cover the parts of the building in detail; starting with the exterior:

External works: A note should be made of the type and condition of paths and pavings and also the condition of gates, fences, railings and boundary walls.

Foundations: These cannot usually be inspected but note should be taken of any settlement cracks, signs of movement around openings, or walls out of plumb, which may indicate a failure in the foundations.

Walls: The report should describe the condition of the fabric in the case of stonework, or brickwork, and whether or not the material has been affected by frost. The state of the pointing should also be noted. The wall thickness should be measured and position and type of damp-proof course should be described. The age of the building, the thickness of the wall, and the pattern of the bonding will help to indicate if the wall is of cavity construction. Special note should be taken of the condition of parapets and copings and of the construction at the head and sill of openings.

Glazing: The type of glass used should be noted and the method of glazing of the frames whether by putty or bead. The number and size of broken panes should be counted.

Windows: A note of the condition of the timber or metal frames and a comment on the catches and method of opening.

Doors: A description of the type of door, its condition and furniture.

Roof coverings: A description of the materials, and a note of the number of broken units in the case of tiles or slates. This item should also include a note on the type and condition of the flashings to the roof.

Gutters and rainwater pipes: Detailed note of the condition of gutters and downpipes and the number of lengths which require replacing.

Paintwork: The state of the paintwork should be noted, and an estimate should be made of the period of time elapsed since the work was carried out.

The third part of the report should deal with the interior; very often this section will detail each room separately under the following headings, starting on the ground floor:

Floors: The material used and a description of its condition, and in the case of timber floors whether or not signs of dry rot or woodworm are apparent. In particular the adequacy of underfloor ventilation should be checked.

Walls: This should deal with the condition of the walls (excluding decoration) on state of plasterwork, and signs of termite attack in panelling.

Ceilings: As for walls.

Decorative order: This should be described separately from the notes on the structure.

After the rooms have been described, an inspection of the roof space should be made.

Roof space: The roof construction should be examined for any signs of fungus or termite attack as for floors. If either of these are found the extent of the damage should be carefully set down. If parts of the floor or roof space cannot be examined, this should be stated and the reason given (i.e. limited headroom, or difficulty of access). It follows that if parts of the structure cannot be examined then no guarantee can be given in the report that the structure is free from defects. However, a reasonable assessment can be made of the likelihood of serious trouble, provided that the surveyor is experienced, and so the client can be advised accordingly.

The fourth part of the report should deal with the services: plumbing, heating, electrical installation and drainage, and the surveyor should call in a specialist if he finds that he cannot properly advise the client on any point. Particular note should be made of defects which are likely to increase the risk of fire or accident, such as loose gas connections or poor wiring. The final part of the report should make recommendations for repairs, and this should be subdivided into items requiring immediate attention and which if not attended to would threaten the safety of the building or its occupants, and items which ought to receive attention as soon as possible.

Where defects are noted, their cause should be listed wherever possible, i.e. 'Dry rot in floor timbers due to lack of ventilation of underfloor space combined with rising damp in external walls' or 'deterioration of plaster under window sills due to open joints in terracotta sills'.

The detailed specification for any repairs to be carried out does not form part of a structural report, but the client will almost certainly ask for this afterwards together with an estimate of the likely cost of the work.

Although as it is obviously not possible to list all the defects that are likely to be found in older buildings, the following comments of the more common faults may be useful as a basis for further inquiry:

Dampness and the various methods of inserting damp-proof courses are dealt with in some detail, beginning on page 201. In buildings which already have a damp course, and the source of trouble is driving rain, it may be possible to effect a cure by the use of a bitumen-impregnated dovetailed sheet which can be nailed to the inside of the old damp walls after the original plaster, if any, has been removed. The sheet acts as a key for new plaster and must be nailed to the wall without timber plugs and using rustproof nails. An alternative is to paint the walls with a rubber-bitumen emulsion which can be 'blinded' (i.e. surfaced) by sand while still wet, to act as a key for the plaster. Both treatments should be carried out from floor to ceiling. These treatments are effective only provided the wall can dry out to the exterior, and if either of these methods are used on walls which have no damp course, then they can only be expected to be a temporary expedient since the rising dampness in severe cases will eventually appear above the waterproofing. In order to prevent dampness being too easily absorbed through an old and very porous brick or stone wall from the outside, one of the many very good silicone treatments can be used. The walls must first be brushed down to remove the loose surface and then the joints raked out ready for repointing. The silicone solution should be applied by a large brush rather than a spray and the brush should be dipped in the solution and laid on the surface so that the liquid runs down and soaks in. There should be no attempt at brushing the liquid into the surface. Pointing should be carried out after the silicone solution has been applied since the liquid will inhibit the drying out of the mortar. When repointing old brickwork or stonework it is a mistake to specify a dense cement mortar since this will crack away from the porous material. The mix should be designed to approach the same porosity as the walling; a guide to a suitable mix would be 2 parts non-hydraulic lime to 9 parts sand, gauged with 1 part cement to 9 parts of the lime—sand mix. Repointing is usually required first in the perpendicular joints since these are not always properly bedded being 'buttered' on the back edge only, when the wall is first built. It is not practical to repoint the perpends without raking out the bed joints and so the whole wall should be repointed. The best time of the year to carry this out is mid-autumn after the hot sun of summer, which would dry out the thin strips of mortar too quickly, and before the frosts which will cause the pointing to crack.

Where smooth cement rendering has deteriorated it will be found that the surface has 'crazed', forming a network of hair cracks. It is not a good practice to use silicones to seal the wall since the material will not permanently 'bridge the gap'. The best treatment is to resurface the wall with a more heavily textured render which will tend to hold the water on the surface and allow evaporation to take place more easily.

Valley gutters and parapet gutters in older buildings are a likely source of trouble. They are usually lead-lined, and where, in an industrial atmosphere, the lead has worn thin, pin-holes appear which are very difficult to see and which cause a slow seepage of water that can remain for a long time undetected. Soldering, although relatively inexpensive, is not an effective method of repair since a crack forms quite soon at the edge of the solder. On important conservation work the old sheets should be removed completely and new sheets inserted, being jointed to the old by the technique of lead-burning. Alternatively preformed fibreglass sheets can be used. These are usually grey in colour.

A small leak remains undiscovered because the water soaks into the roof timbers and wall plates and does not drop on to the ceiling where it would cause stains which would be easily seen. Thus the roof timbers in the vicinity of a damaged gutter should always be carefully examined for decay.

When slates (or tiles) have to be repaired the last row cannot be nailed since the head of the slate must be pushed up under the two upper layers of slate. The traditional strip of lead nailed to the lath underneath the slate, and then bent over the lower edge to secure it, is not effective since the lead will straighten out under the effect of snow and even high winds and will allow the slate to drop down. A strip of copper similarly used is better, but stout copper wire nailed to the lath as before and then turned up over the bottom edge in the form of a coil spring, is the best method.

Leaks from flat roofing which is incorrectly laid can be very troublesome. The cause is usually due to excessive thermal movement. So every precaution should be taken in the design and detailing as well as in the workmanship and materials used to ensure a satisfactory covering. Specialist labour and techniques are usually required to lay the covering as with built-up felt, asphalt, copper and aluminium, and so specialist labour is required to make an effective repair.

Plastics in thick liquid form are available for application to repair leaks in flat roofs. The liquid is applied in two or more coats which form an impervious membrane which will stretch with thermal movement to bridge hairline cracks. These plastics, however, should be considered in a similar way to a protective paint, since the membrane will require renewal every three or four years.

One of the most persistent complaints from building owners is that of chimneys or heating appliances, which send smoke or fumes back into the room. This is not always an easy matter to solve because of the numerous combinations of circumstances which can cause the smoking back. The basic causes, however, provided that the flue is long enough and the chimney well built and properly lined, are air starvation from within the room due to excessive zeal in preventing draughts, and external pressures causing downdraughts. These external downdraughts can be caused by high trees in the vicinity, the general lie of the land, or wind deflection from adjacent high buildings. In order to prevent smoking back caused by air starvation, it is a good plan to increase the effectiveness of the air current and to reduce the amount required, by fitting a throat-restricting device. Where external air pressure is causing downdraught, the chimney should be built up to raise the stack above the pressure area wherever possible. Where this cannot be done, a special chimney pot may be fitted which causes the downdraught to bypass the top of the flue and create an uplift. There are several types of special chimney pots, one of the best known being in the shape of a letter H. The draught passing down the vertical leg of the pot causes the air to be drawn quickly across the horizontal portion and so draws the smoke up the chimney. This type of pot is very effective but is rather large and weighs over 50 kg, so care must be taken to secure it adequately. Leaking joints in the chimney flaunching or in the brickwork can create downdraughts since the cold air will cool the flue.

CHAPTER THIRTEEN

Building Alterations and Repairs

More money is spent each year on repairing and altering old buildings than on building new ones. Nearly all small and many medium-sized building firms would be better described as building repairers, and for this type of work much extra knowledge is needed beyond that of ordinary building construction. Practical experience is particularly needed, because many judgments, such as how stable an old wall may be, will have to depend on previous experience of similar cases, even sometimes the experience of failures.

Dampness in buildings
One of the most usual defects to be found in an old building is dampness. In trying to trace the source it will probably be found to have come from one or more causes. The most likely ones are briefly listed below:
  (1) Defective construction, such as broken rainwater pipes, leaking or choked gutters or missing roof tiles.
  (2) Defective flashings at chimneys or at the junction of high and low building.
  (3) Defective leadwork in valley gutters.
  (4) Wind-driven rain passing right through the thickness of a solid wall.
  (5) Damp rising from the ground, either because there is no damp course or because the existing damp course has been bridged by the earth outside being banked up to form a flower bed or rockery above the damp-course level. It is also possible for an existing damp course to have been bridged by cement rendering applied to an old wall and being carried down below the damp-course level and coming into contact with the wet topsoil. Methods of inserting damp courses into existing walls to prevent rising damp are discussed later. Note should be taken of the fact that it is possible for rising dampness which has been present over a long period of years to have carried up into the wall

certain hygroscopic salts. These salts will in turn, even after the rising dampness has been cured, absorb moisture out of the atmosphere and this will appear again as dampness on the surface. When it is ascertained that the persisting dampness is attributable to this deliquescent action of the hygroscopic salts then a neutralizing agent can be applied.

(6) Blocked ventilation gratings in external walls.
(7) Leaking water taps and wastepipes for sinks and wash basins.
(8) Internal condensation. Where humidity is high, the moisture droplets of condensation will be found on all cold surfaces. This is due to the fact that the warmer the air, the more water vapour it can hold so that when moist, warm air cools, either as it passes through a wall or when it comes into contact with a cold surface, the fall in temperature causes the vapour to condense into drops of water. On porous surfaces such as soft plaster the condensation will cause mould growth, and on hard impervious surfaces the water will run down or drip off. The remedy for condensation is to provide continuous background warmth so that the air retains the moisture vapour at a temperature above the dew point. It is also necessary to provide adequate ventilation, sometimes by a mechanical extraction system, in order to remove the moist air and to provide adequate thermal insulation on the interior surface of a wall so that the surface temperature remains above the dew point.

Dry rot

Dry rot is an infectious disease of timber. Once it is established it spreads rapidly not only in the infected building, but to adjacent properties as well.

The manifestation of dry rot is fungus (*Merulius lacrymans*) which denatures timber, particularly softwood, drawing away its moisture and strength. After the attack the timber is left dry and rotten, and thus the term 'dry rot'.

The disease will start in any damp unventilated place where timber is to be found. This is commonly under old suspended timber floors, in skirtings and edges of floors near external walls, and in roofs near leaking gutters. Very tiny dust-like reddish-brown spores which are carried through the air from an established infection will start a new outbreak where the conditions are right. The fungus starts by producing a cottonwool-like mass, with patches of yellow, and spreads through the timber by putting out root-like grey strands which infiltrate also through mortar cracks in brickwork and behind plasterwork. These strands carry moisture from the source of dampness so that the fungus will spread to other woodwork which may originally be quite dry and in a well-ventilated position. Finally, the fungus grows a large white and yellow fruity body which liberates from its surface millions of the reddish-brown spores each of which will germinate and start a fresh outbreak where conditions are favourable. A smell similar to that of mushrooms is always associated with an outbreak of dry rot.

It is thus most important that all evidence of dampness is investigated and the cause put right. A collapsed floor board, irregularities or waviness appearing on the surfaces of skirtings, window linings, or wall panelling in the vicinity of dampness will most probably indicate that an outbreak of dry rot is established. When the outbreak has been discovered it is important that the whole area both in the vicinity of the attack and for at least 1 m beyond is cleared of infection. All outbreaks should be investigated by an experienced surveyor who will advise on the amount and extent of the repairs. The work will consist of cutting out all decayed timbers, hacking away affected plasterwork and removing timber skirtings and linings. All the decayed and affected timbers must be removed from the building immediately and burned, and then the walls in the vicinity of the outbreak must be treated to kill the fungus. This was traditionally carried out by heating the surface with a blowlamp but this is a somewhat dangerous technique and does not in fact penetrate deep into the wall. The present-day treatment would be to irrigate the walls by boring holes into the brickwork at the perimeter of the outbreak and pumping in fungicidal liquid under pressure to isolate the outbreak. All existing timbers surrounding the infected area should be treated by brushing with at least two coats of fungicidal fluid. Also all new timbers used in repair should be treated similarly. There are many very good proprietary fungicides on the market and many firms operate an advisory service. The fungicide should conform to the requirements of B.S. 1282. As an additional precaution new plasterwork could have a floating coat of special zinc oxychloride plaster which inhibits the spread of fungus.

Wet rot
There are many types of fungus that will attack timber and in particular the dry rot fungus should not be confused with an outbreak of wet rot, caused by *Coniophora cerebella* or 'cellar fungus'. It is important that the type of fungus is correctly identified since the treatment technique

depends upon the nature of the attack, and the knowledge of an experienced specialist surveyor is necessary.

Wet rot grows in very damp conditions and requires twice as much moisture as dry rot before the spores can germinate. Provided that the cource of dampness can be removed and the timber dried out, no further attack will take place in timbers not already affected. The treatment will consist of removing all the diseased timbers and replacing them with new dry timber treated with fungicide.

Woodworm

Timber is attacked by several species of woodboring insects, whose larvae eat their way through the timber emerging as beetles and leaving the characteristic hole 2 mm to 6 mm in diameter in the surface of the timber, depending upon the species. The beetles then lay eggs on the timber surface which in turn hatch out into grubs and the cycle starts once again. The most common inseect is the Common Furniture Beetle, but depending on the age, condition, and situation of the structure attack may also be expected from the House Longhorn Beetle, the Powder Post Beetle and the Death Watch Beetle. Where the attack has been so severe as to destroy the nature of the timber, the affected timbers must be removed and repairs using new treated timber must be carried out.

Treatment of the remaining affected timbers would take the form of spraying with the correct insecticide. Special spray apparatus must be used so that the correct type and amount of liquid can be pressure-injected and the advice of a specialist surveyor should be sought.

The liquid acts by penetrating the timber and poisoning the larvae and also provides a penetration barrier near the surface to safeguard the timber from further attack from outside.

Attack by woodworm, in particular the House Longhorn Beetle, is so severe in certain areas that the Building Regulations require that all structural softwood is given preservative protection before it is used in these areas.

Protected timber

It will be seen that the damage done by the fungus and termite attack in timber is a very serious matter affecting the whole economics of building. It is therefore a good thing when considering the timber specification, not only for repairs but for all structural timbers in new work, to ensure that the timbers are pressure-impregnated with liquid preservative to give protection against woodworm and dry rot. There

BUILDING ALTERATIONS AND REPAIRS · 197

are many reputable proprietary systems available and timber merchants now often carry stocks of treated timber. For an extra cost, timber can also be pressure-impregnated to make it resistant to fire.

Shoring

Shoring has to be undertaken sooner or later by every urban builder, and though the usual methods adopted can be described, it is essential that each case be studied very carefully and a clear picture seen of how the loads are distributed and what is likely to happen when the shoring takes up the load. This must be held in the mind of the person responsible for the work and explained to the workmen engaged on it.

When part of a building needs holding up, supporting vertically, as when the lower part of a wall has to be rebuilt or is taken away to form a shop window, a system of dead shores is needed. When part of a building, usually a wall, is likely to move sideways due to lack of lateral support, a system of raking or flying shores is needed.

Diagram 164 shows a system of dead shores supporting the front wall of a terrace house, so that a large opening can be cut and a long structural steel beam inserted to carry the old wall over. Beams called 'needles' are placed in holes cut in the wall in positions arranged so that the needles support the piers. These needles can be wood or steel beams

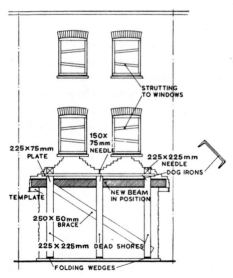

Diagram 164   Dead shoring: elevation of system

and they should be designed for the load of brickwork and floor each will have to carry. Baulks of timber in large sizes such as 225 × 225 mm are kept by most builders for shoring and needles of this kind. Those about 2 m long are suitable for supporting piers between windows, as in the diagram. The needles are supported at the ends by wooden plates, supported in turn by the dead shores; in this example these are also 225 mm × 225 mm. Of course, each dead shore should come under the end of a needle, but this is not always possible, and in that case the plate may have to be stronger, as it will then be acting as a beam.

The dead shores are to take the load down to the ground or other firm foundation. A trench can be dug to a depth where the soil is firm and the lower plate or sill can be laid along this. Sometimes the made-up soil under paving is sound enough. Inside the building it is necessary to go down to the surface concrete. The dead shores should stand away from the wall to allow working space; this distance affects the size of the needle, and if much working space is needed a stronger needle will have to be used.

After these members are in position they are tightened up by folding wedges at the base of the dead shores and are secured in position by dog-irons and by braces nailed across the sides of the shores. When these are all fixed the rest of the brickwork can be cut away. If the brickwork is in soft mortar, it will only corbel to the line of joints as sketched in diagram 164 and temporary props may be necessary to keep up the sills; alternatively, it might be better to remove the sills and refix them afterwards. If the mortar is good, the brickwork under the sills will 'hang up'. After the beam is inserted it is important that the brickwork is built up over it to meet the old brickwork as closely as possible; this is called 'pinning up' (diagram 165).

It is difficult to ensure that some slight movement will not occur in shoring operations of this kind, and small cracks may develop and damage arches if windows are not strutted as shown in diagram 164. A 100 × 75 mm plate stands on the sill and a member is shaped to the arch for the head. Posts are cut in between them, and these posts are wedged tight to the reveals with struts cut to a tight fit and driven in obliquely as shown.

Diagram 166 shows how additional strutting is necessary to support floor joists that bear on the wall to be shored. 100 × 75 mm struts at about 1·5 m centres support a plate under the joists and transfer the weight to the floor below. To relieve the wall this strutting can go on up and support all the floors above; in this case blockings are needed between the surface concrete and the lowest floor joists.

BUILDING ALTERATIONS AND REPAIRS · 199

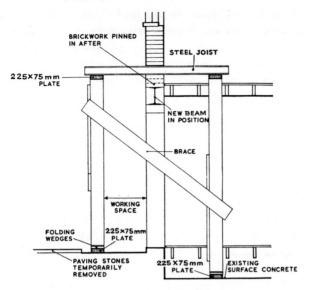

Diagram 165   Dead shoring: section

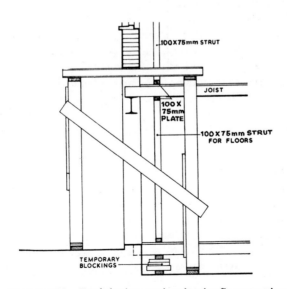

Diagram 166   Dead shoring: section showing floor strutting

200 · BUILDING TECHNIQUES

A raking shore is shown in diagram 167. The aim is to provide an inclined support directly on the centre line of thrust from each floor, from the wall plate or, where the joists are parallel with the wall, at the junctions of the centre lines of the floor and wall. A low inclination may in some cases be better (the purpose only being to prevent sideways movement), but in practice it is difficult to get long enough shores and too much space may be taken up. An angle of 70 degrees with the horizontal for the outer shore is the maximum. To support the

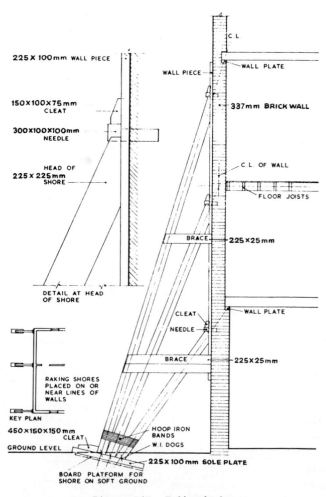

Diagram 167   Raking shoring

brickwork a 225 x 100 mm wall piece is fixed, the needles being fixed through this into holes cut in the wall and the head of the shore is notched to fit under this needle.

A firm base is essential; in soft ground a board platform is formed on which the sole plate is laid. The lowest shore is fixed first, the others being cut to fit tight behind this and levered into position with a crowbar. Braces, cleats and hoop-iron binding strengthen the frame. Note that no folding wedges are used. A proper tight fit should be obtained by accurate setting out; wedges should never be used. It is dangerous to try to push back a wall which is overhanging by wedging a system of raking shores.

Diagram 168 shows a system of flying shores. The purpose is the same as for raking shores; it leaves the ground quite free for building operations, but can only be used in cases such as rebuilding a terrace house where walls opposite each other have to be strutted.

The arrangement of the members explains itself when the principle of the raking shore is understood, but folding wedges have to be introduced to make erection easier.

Underpinning

When a new building is built against an old one, and has its foundations at a lower level, the placing of these foundations presents a problem. Trenches cut for the new foundations would undermine the old wall and it might sink in spite of efficient shoring. For this reason it is advisable to underpin the foundations of the old wall first. This involves digging a series of holes under the old foundations to the depth of the new foundations and filling up or underpinning with concrete. By doing this several times in the order shown in diagram 169 no great length of wall is left unsupported for long and the result is a better foundation at lower depth, allowing the whole of the building site adjoining to be excavated for a basement, as shown in the figure.

The same method of underpinning in short lengths is used wherever a horizontal cut is made in a wall, as in the case of inserting a new damp-proof course. Many old buildings have no damp course, or the original one has become defective. New damp-proof courses can be formed by cutting out say lengths of 1 m at a time and inserting two courses of slates, or bituminous felt or whatever is desired, and 'pinning up' in brickwork over. The best material to use for the damp-proof course in this case is probably blue Staffordshire bricks, as there is then no difficulty in getting in the last course of bricks.

202 · BUILDING TECHNIQUES

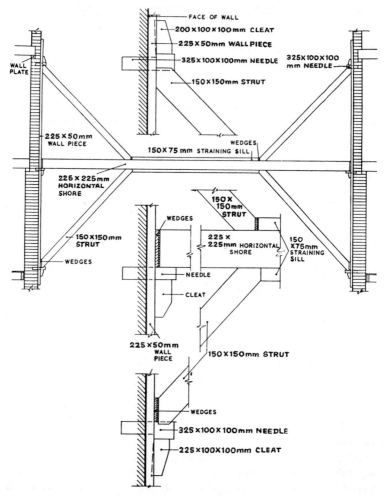

Diagram 168   Flying shore details

Underpinning in short lengths is necessary sometimes to cure a settlement, but the cause of the settlement should be investigated thoroughly before any underpinning is done. The most frequent cause of a settlement in a wall is not usually the great weight of the wall but the insufficient depth of the foundations. This has been discussed in chapter 3, but additional causes may be (*a*) defective drains, which allow parts of the soil to become plastic and squeeze out under the weight, or (*b*) tree roots, which can be strong enough to lift up parts of a foundation. A section of underpinned wall is shown in diagram 170.

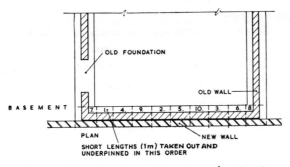

Diagram 169   Order for underpinning in short lengths

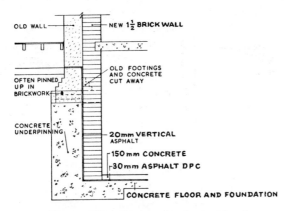

Diagram 170   Section of underpinned wall

'Saw-cut' D.P.C.

An alternative method of inserting a damp-proof course in an existing wall has now been developed which is less costly than underpinning. A special saw is used having a blade of mild steel about 3 mm thick and with teeth tipped with tungsten carbide. The saw may be provided with handles and guards for use by two men (one either side of the wall), but will preferably be power-driven and may also be provided with a cradle on wheels to make it self-supporting from the ground.

A damp-proof course is inserted immediately the cut is made and normally in lengths of about 600 mm. The best membrane to use for this work is copper sheet and provided that no snags are encountered, the cutting and inserting of the metal sheet can proceed at the rate of just over one metre per hour in one brick thick walls. This method is not practicable in unsound brickwork since the bricks are loosened and fall as the saw advances.

Electro-osmotic D.P.C.

It is now possible to cure rising damp in solid walls by the use of an electro-osmotic process patented by the Rentokil Laboratories Ltd. It can be shown by experiment that damp walls are electrically charged in relation to the ground on which they stand and as evaporation losses from the wall are replenished by rising damp, the wall acts as if it were an accumulator on permanent trickle charge. This charged state will, in fact, accelerate the flow of water through the construction.

The object of the installation is to provide a specially designed circuit of very low electrical resistance in order to break down the electrical charge originally induced in the wall by the capillary movement of the rising damp. When the charge is broken down, the dampness is no longer able to pass through the wall by capillary attraction, and so the wall begins to dry out and then will remain dry.

The circuit consists of wall and earth electrodes connected together. The wall electrode is in the form of a continuous strip of pure copper fixed into each length of the damp wall. This strip is placed in a raked-out joint in the brickwork and looped into holes drilled at specific distances and depths into the thickness of the wall. The joint is repointed with special mortar when the electrode is in position. The details and positioning of the electrode bank is a matter for the specialist surveyor. To form the earth electrode, copper-covered steel rods are driven into the ground at a given distance apart and to a predetermined depth. This distance and depth is calculated by measurement of the electrical resistivity of the soil and will vary between 10 to 12 m apart and between 4 to 6 m deep. The rods are protected against corrosion in the ground by a special chemically inert plastic compound. The wall electrode bank and the earth electrodes are connected by means of a stranded copper wire protected by PVC sheathing. Diagram 171 illustrates the basic layout of the system.

The electro-osmotic process has the advantage that it is not so difficult to treat thick walls as with the sawcut method.

Chemical injection damp proofing systems

Methods have been developed by specialist contractors of injecting under pressure solutions of silicone resins (or other suitable chemicals) into existing solid brick or stone walls to provide a barrier against rising damp. The wall is drilled at pre-determined intervals and to a pre-determined depth dependent upon the thickness of the wall to be treated. Preferably the holes are drilled horizontally into a mortar bed joint. The injection of the chemical solution is usually carried out under

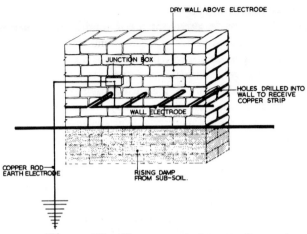

Diagram 171   Electro-osmotic damp proofing

pressure. The drilled holes are afterwards plugged with cement sand mortar, and where the wall has been previously plastered internally this must be re-plastered using a special mix. The drillings for the damp proof course must extend at least 150 mm above ground level and must be positioned below the timber joists and/or wall plates in the case of suspended timber floors.

Drying out
After a damp wall has been provided with a new damp-proof course, the wall must be given time to dry out before redecoration is contemplated. This drying out may take a long time and it would be wise, for instance, to allow up to 9 months for a one brick thick wall (1 month per 25 mm of thickness).

Replastering
If an old wall has been damp for a number of years, it is probable that the plaster will have perished and will have to be removed. The new plaster should not be a strong mix, nor should it be trowelled to a hard finish since a polished surface will act as a vapour barrier and cause condensation to form. Not many householders are willing to wait long periods until the walls dry out, so when the new plaster appears dry to the touch it can be covered by a permeable emulsion paint or distemper which will allow the drying process to be completed without damage to the paint film.

Building measurement

When a building is to be altered or major repairs undertaken then a measured survey must be carried out. The extent and complexity of the alterations, or repairs, proposed will determine the scale to which the measured drawing will eventually be drawn out and this in turn will indicate the detail and method of measuring required on site.

A measured drawing is the best way to record buildings of architectural and historic interest. A knowledge of construction is required so that the building may be assessed and measurements taken to reveal the structure, since a traditional building is a structural skeleton clothed in plasterwork and panelling. Certain partition walls will be non-load-bearing and it is important to be able to discriminate between the parts so that the structural skeleton is revealed. The direction of floorboarding, which indicates the opposing span of the joists, should be noted, the lines of nails, where visible, show the spacing of the joists. Pattern staining in the plasterwork will reveal the lines of the supports to wall panelling and the centres of ceiling joists. Check measurements must be taken to allow for the thickness of plasterwork so that the structural thickness of the wall can be shown, bearing in mind that in older buildings the plaster may be up to 38 mm thick and that the bricks will probably be larger than the standard in use today. Where the building has a structural framework, as in the mediaeval timber building, or has cast-iron columns and beams as in the nineteenth-century factory buildings, the lines of the framing should form a basis for the measurements.

A survey should be divided into two distinct parts, drawing first and then measuring when the drawing is complete. Thus the building is first sketched, then the measurements are added and then finally the measured drawing is prepared from the sketches. The sketches should be freehand and should delineate the building in plan, elevation and section; the size of the sketches should be as near as possible to the intended scale drawing of the building. Although it is not practicable to draw the first sketches to scale, very great care should be taken to draw the building to the correct proportions, and to add all the lines which indicate the points of measuring. It is not good technique to prepare rough, badly proportioned sketches and rely on memory if the measurements do not appear to check with the sketch. The aim in preparing the measured sketches should be to make them so accurate and clear that they could be plotted and drawn out by someone else.

The equipment required is a measuring tape, a measuring rod, and a sketch pad. The tape may be either of synthetic material — usually woven fibreglass with a coating of PVC, or alternatively of thin flexible steel strip. The non-metal tape is adequate for most work and is easy to handle but tends to stretch or shrink in use, and so for more accurate work flexible steel tape should be used. A steel tape is more expensive than its plastic counterpart and is more difficult to use and can fairly easily snap if bent in the wrong direction. The size of the sketch pad used depends on the detail required on the drawing but an A3 (297 x 420 mm) sheet is the largest that can be conveniently handled. Where the work permits, the A4 sheet (210 x 297 mm) would be a good size to use since the measured notes could then be filed as part of the job record. There are many sketch pads on the market, some utilizing a system of translucent paper on a squared backing which forms a grid for the drawing. Many surveyors, however, use a thin plywood sheet as backing and attach loose sheets of drawing paper by clips to act as a pad. The measuring tape will be wound on a spool either in a plastic or leather case. Tapes are obtainable to B.S. 4484 usually 10 m, 20 m or 30 m long. Retractable steel rules are also very popular and are made in lengths of 1, 2, 3 or 5 m. On metric tapes the principal (first-order) marks are at 100 mm intervals with secondary (second-order) marks at 10 mm (1 cm) intervals. The millimetre marks are also indicated, but sometimes only in the first metre of the tape. Folding boxwood rules are also popular and are available in 1 and 2 m lengths when extended. Care should be taken when winding the tape so that it does not twist into the casing. Every care should be taken to keep the tapes off the ground, but if a tape gets wet and muddy in use it should always be cleaned off and in the case of a metal tape wiped with an oily rag when the work is finished.

Diagrams 172, 173, 174, 175 and 176 show the techniques used in sketching, and some of the problems to be solved in various types of building. In order to get the correct proportion of the building, it is a good plan to use an imaginary square as the unit of proportion. Squared paper can be used as the survey sheet but there is a tendency for the lines to confuse the sketch.

Wherever possible, measurements should be taken from one point on a line picking up each feature of the building so that the totals on each line are cumulative. This is known as 'running measurement'. Alternatively, where this is not possible, each measurement can be noted separately. The running measurements are to be preferred, since an error on one measurement along the line does not invalidate the rest of the work, nor does it make the total length incorrect. It will be seen that the best system therefore requires that two people should carry

208 · BUILDING TECHNIQUES

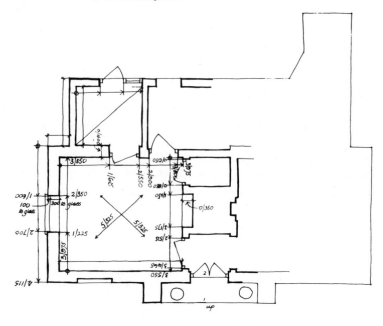

NOTES

1. CHECK IF ROOMS ARE SQUARE BY MEASURING DIAGONALS
2. READ TAPE FROM LEFT TO RIGHT
3. DIMENSIONS SHOULD FOLLOW CLOCKWISE AROUND THE ROOM
4. NOTE THICKNESS OF WALLS AT OPENINGS
5. CIRCLE DENOTES START OF A SERIES OF RUNNING DIMENSIONS AND ARROW HEAD DENOTES END
6. EXTENT OF SINGLE DIMENSIONS NOTED BY DIAGONAL LINES

Diagram 172   Measured work – typical plan

out a survey, one to hold the tape and note the measurements and the other to read the tape. A team of three is better still, with the junior holding the blank or 'dozy' end of the tape, and leaving one person free to concentrate on the drawing. The run of measurements should always read from left to right. It is also helpful to orientate the drawing sheet along the line of the measurements so that there can be no confusion as to which point the figures relate. Running measurements are indicated by a dot in a circle at the start of the line and arrowheads at each cumulative point. Single measurements can be noted by small diagonal lines at each side. It is important to distinguish between the two on the sketch since often in difficult situations the two methods have to be used together and mistakes can easily arise when plotting the

BUILDING ALTERATIONS AND REPAIRS · 209

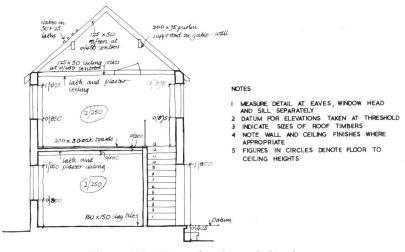

Diagram 173 Measured work – typical section

measurements. The degree of accuracy necessary in measuring depends upon the scale to which the drawing is to be plotted and the intricacy of the work. For checking mouldings on timber or plaster the work may have to be accurate to 1 millimetre, but for normal work to be plotted to say 1 : 50, measurements to the nearest 10 mm (1 cm) are satisfactory since this is the limit of accuracy that can be attained on drawing out. For plotting to 1 : 100 or 1 : 50, any timber architraves

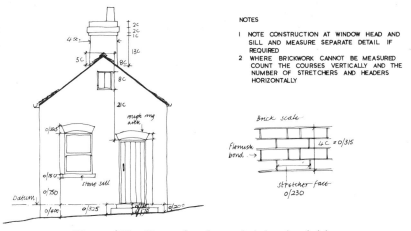

Diagram 174 Measured work – typical elevation: brick

210 · BUILDING TECHNIQUES

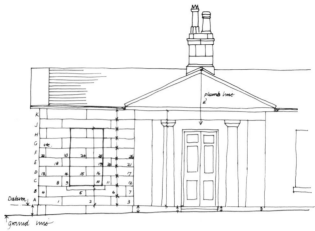

NOTES

1. WHERE STONEWORK IS NOT IN COURSES EACH STONE SHOULD BE INDEXED AND MEASURED
2. WHERE STONEWORK IS IN COURSES EACH COURSE CAN BE MEASURED AND THE WIDTH OF EACH STONE RECORDED AS IN THE EXAMPLE.

Diagram 175   Measured work – typical elevation: stone

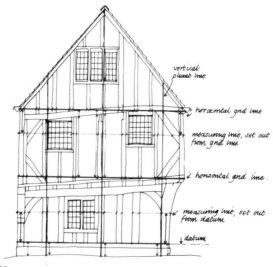

NOTES

1. ASSUME DATUM
2. GRID LINES SET OUT ABOVE DATUM WITH ALL VERTICALS PLUMB TO IT.

Diagram 176   Measured work – typical elevation: framed construction and mouldings are ignored and the measurements taken to the inside face of the timber window or door lining to give the net opening width. It is often necessary to draw the detail mouldings to a larger scale, and measure separately. Check diagonals should be taken to see if a room is square on plan. Sections are the most difficult part of the measuring but it is necessary for an accurate section to be measured since this is an essential part of the information required when a building is to be altered. Running measurements are not often possible in measuring sections and so special care should be taken to see that the totals check. It is useful to note the height of each room separately. The floor thickness can usually be measured at the stairs.

For elevations in brickwork the brick scale is noted by measuring the height of, say, four courses (and four joints) and then the elevation can be measured by counting the number of courses and relating them to the brick scale when the elevation is drawn out. This technique is particularly useful for noting parts of the building that are inaccessible, such as chimneys. Widths in such situations can also be measured by counting the number of stretchers (225 mm face) and headers (112 mm face). The rise of all arches over windows and doors should be noted; often it is possible to get necessary elevational detail from upstairs windows.

Elevations in stonework present a more difficult problem, and where the stonework is not in courses, for example, in rubble walling, each stone should be indexed and measured where the quality of restoration work demands, this amount of detail. In ashlar work, which will be coursed, the height of each course can be measured and the width only of each stone recorded as in diagram 175.

For old timber-framed buildings special techniques have to be used. The elevation must be provided with grid lines and this is done by the use of a plumb line for the vertical grid lines, and a spirit level for the horizontal grid lines. The salient points on the grid can be indicated in a fine chalk line on the building and the framework picked up from this as shown in diagram 176.

The diagrams indicate the methods used in measured work and are not intended as examples of site sketches.

APPENDIX:
Further Sources of Information

The student will find the following additional sources of information useful in keeping up to date with current practice.

British Standards

Most raw materials, components and fittings used in the building industry are covered by British Standards. (usually abbreviated to B.S.). These Standards are produced under the supervision of special committees formed by representatives of the Professional and Technical Institutions concerned with building and they give information relevant to the performance of a particular material or article. Wherever possible, tests have been devised to check this performance and these tests are often described in the relevant Standard. Thus when a British Standard is quoted, the Contractor will order accordingly, and the manufacturer or supplier will guarantee the goods to the Standard. This means that site testing is not necessary and that long and perhaps ambiguous specification of materials for each contract can be avoided. Where possible, materials and components must be stamped by a code or distinguishing mark to show that they conform to the appropriate Specification.

A full list of British Standards published by the British Standards Institution numbers over six thousand, covering industries other than building, and the student should always take the trouble to study the British Standards relevant to his particular work and should avoid quoting the Standard unless he is familiar with the implication of its contents.

The Standards are, of course, revised when necessary and the date of revision is incorporated in the title, thus; British Standard 449: 1970 *The Use of Structural Steel in Building* is the latest current edition. This particular Standard was first issued in 1932, revised in 1937, 1948, 1959 and again in 1969, and so it is important to make sure that the latest revision is used for reference.

British Standard Codes of Practice
There are also Codes of Practice (B.S.C.P.) issued by the British Standards Institution on behalf of various technical committees: Building; Installations; Services (this includes drainage; water and gas supply; mechanical ventilation; fire-fighting installations; and electrical installation); Civil Engineering and Mechanical Engineering. The Codes set out the best current practice in the light of present-day experience and knowledge.

Building Research Establishment (B.R.E.) Digests
These are leaflets issued each month which review current materials and techniques and assess their effectiveness.

Agrément Certicates
The Agrément Board was set up in Britain in 1966, based on experience in France, to develop test procedures and methods of assessment (published as MOATS) in order to be able to evaluate new materials, products and techniques which manufacturers intend to put on to the market.

The Board grant Certificates to manufacturers when their products have satisfactorily passed the agreed tests. Subsequent to the granting of Certificates, strict quality control must be maintained and the Certificate must be renewed after three years.

Government publications
There are many leaflets and booklets published by government departments which give technical information on special aspects of building as indicated by the following brief list:

> *Advisory Leaflets:* These leaflets contain information of a practical nature, covering a wide range, to include setting out, concreting, fixing windows, etc., and are primarily intended for the workman on the site.
>
> *Safety, Health and Welfare Booklets:* This series is intended for other industries, but the relevant booklets cover scaffolding and safety in excavations.

Trade literature
Many trade associations and manufacturing firms produce excellent descriptive literature, often in attractive bindings, with technical information on the products and techniques concerned. The student

will undoubtedly find these useful but should always be careful to discriminate between the technical and the advertising matter.

R.I.B.A. Product Data
Royal Institute of British Architects Services Ltd produce library data sheets giving information on building products and services. These give product information under standard headings in a standard order. The data sheets are based on information given by manufacturers and the standard format helps in the comparison and identification of products. As a supplement to the data sheets, *Construction Notes* give information on new techniques and monitor failures in building, and *Practice Notes* deal with matters arising in architectural practice.

Classification of information
It will be seen that because of the very wide range of subjects embraced by building and the mass of literature appropriate to each activity, a special means of classification of information is necessary and a system known as CI/SfB has now been adopted for this purpose. The well-known UDC (Universal Decimal Classification) system, which is used in libraries for the general classification of books, is essential as a primary classification but is not so appropriate to the very detailed breakdown of information required over the wide range of specialist subjects covered by building. The CI/SfB system is thus designed specially for those engaged in building, both for the filing of technical information and for the referencing of working drawings. The standard size of technical literature for classification is recommended to be 297 mm x 210 mm, to B.S. 1311, which is known as the A4 sheet. Most technical literature now has the CI/SfB classification printed in the top right-hand corner of the front sheet.

With regard to the material contained in this book, the appropriate CI/SfB classification and a brief list of the relevant British Standards and Codes of Practice are given below under chapter headings:

The CI/SfB system uses both numbers and letters as symbols of the main groups of activities which form the building process. The main divisions including Theory and Practice are indicated by capital letters. Parts of a building which have a different functional use, i.e. Functional Elements such as windows, doors, services, etc., are indicated by bracketed numbers.

FURTHER SOURCES OF INFORMATION · 215

A further subdivision relating to building materials is made by the use of lower-case letters.

For a full explanation of the system, see the CI/SfB Construction Indexing Manual published by R.I.B.A. Publications Ltd.

	CI/SfB filing reference	A selection of the relevant British Standards	British Standards Codes of Practice
The site			
Surveying	(A3s)		
Excavation	(11)		
Concrete pipes for drainage		B.S. 1194	
Clayware field drainpipes		B.S. 1196	
External works			
Roadworks	12		
Pavings	(90·22)		
Garden fences and gates	(90·215)		
Tars for road purposes		B.S. 76	
Concrete kerbs		B.S. 340	
Precast concrete flags		B.S. 368	
Granite setts		B.S. 435	
Sandstone kerbs and setts		B.S. 706	
Tarmacadam, crushed rock aggregate		B.S. 802	
Tarmacadam, gravel aggregate		B.S. 1241	
Mastic asphalt for roads		B.S. 1446	
Bitumen macadam		B.S. 1621	
Fine cold asphalt		B.S. 1690	
Bitumen emulsion for roads		B.S. 2542	
Building drainage			B.S.C.P. 301
Load-bearing construction			
Soil mechanics	(L4)		
Brickwork	2771		
Foundations — general	(16)		
Walling — external load-bearing	(21)		
Chimneys, flues and open fires	(21·8)		
Portland cement		B.S. 12	
Unit weights of building materials		B.S. 648	
Materials for d.p.c.'s		B.S. 743	
Methods of testing mineral aggregates and sand		B.S. 812	
Concrete aggregates		B.S. 882	
Building limes		B.S. 890	
Mastic asphalt for tanking and damp-proof courses (limestone aggregate)		B.S. 988	

216 · BUILDING TECHNIQUES

	CI/SfB filing reference	A selection of the relevant British Standards	British Standards Codes of Practice
Clay fluelinings and chimney pots		B.S. 1181	
Sands for mortar		B.S. 1200	
Cast-concrete lintels		B.S. 1239	
Natural stone lintels		B.S. 1240	
Metal wall ties		B.S. 1243	
Batch-type concrete mixers		B.S. 1305	
Methods of testing soil for civil engineering purposes		B.S. 1377	
Methods of testing concrete		B.S. 1881	
Readymix concrete		B.S. 1926	
Precast concrete blocks		B.S. 2028	
Copper sheet and strip		B.S. 2870	
Clay bricks and blocks		B.S. 3921	
Dimensions for special bricks		B.S. 4729	
External rendered finishes			B.S.C.P. 221
Sound insulation and noise reduction			B.S.C.P. 3, Chap. III
Precautions against fire			B.S.C.P. 3, Chap. IV
Durability			B.S.C.P. 3, Chap. IX
Foundations and substructures of non-industrial buildings			B.S.C.P. 101
Structural recommendations for load-bearing walls			B.S.C.P. 111
Brickwork and masonry			B.S.C.P. 121
Walls and partitions of blocks and slabs			B.S.C.P. 122
Flues for domestic appliances burning solid fuel			B.S.C.P. 131
Framed construction			
Structural engineering	(2—)(K)		
Use of structural steel in building		B.S. 449	
Glossary of terms for concrete and reinforced concrete		B.S. 2787	
Structural (weldable) steel		B.S. 4360	
Timber grades for structural used		B.S. 4978	
Structural use of timber in buildings			B.S.C.P. 112
Structural use of normal reinforced concrete in buildings			B.S.C.P. 114
Floors			
Floors — suspended	(23)		
Floors — beds	(13)		
Glossary of terms relating to timber and woodwork		B.S. 565	
Woodblocks for floors		B.S. 1187	

FURTHER SOURCES OF INFORMATION · 217

	CI/SfB filing reference	A selection of the relevant British Standards	British Standards Codes of Practice
Grading and sizing of softwood tongued and grooved flooring		B.S. 1297	
Mastic asphalt		B.S. 988	
Steel for prestressed concrete		B.S. 2691	
Structural use of timber in building			B.S.C.P. 112
Structural use of normal reinforced concrete in building			B.S.C.P. 114
Structural use of prestressed concrete			B.S.C.P. 115
Timber flooring			B.S.C.P. 201
In situ flooring			B.S.C.P. 204

Roofs

Roof structures – framing	(27)		
Finishes to roofs – general	(47)		
Metal flashings	(47)		
Clay plain roofing tiles		B.S. 402	
Concrete plain roofing tiles		B.S. 473	
Fire tests on building materials		B.S. 476	
Asbestos cement rainwater pipes and gutters		B.S. 569	
Roofing slates		B.S. 680	
Asbestos cement slates and sheets		B.S. 690	
Bituminous roofing felts		B.S. 747	
Plain sheet zinc roofing		B.S. 849	
Wood-wool slab		B.S. 1105	
Milled lead-sheet for building		B.S. 1178	
Wrought-copper and zinc rainwater goods		B.S. 1431	
Connectors for timber		B.S. 1579	
Glossary of terms applicable to roof covering		B.S. 2717	
Aluminium rainwater goods		B.S. 2997	
Structural use of timber in building			B.S.C.P. 112
Slating and tiling			B.S.C.P. 142
Sheet roof coverings – copper and lead, etc.			B.S.C.P. 143
Roof coverings			B.S.C.P. 144

Doors

Ironmongery	Xt7		
Doors – construction elements	(32)		
Schedule of sizes for door locks and latches		B.S. 455	
Panelled and glazed wood doors		B.S. 459, Part I	
Flush doors		B.S. 459, Part II	
Fire-check flush doors		B.S. 459, Part III	

218 · BUILDING TECHNIQUES

	CI/SfB filing reference	A selection of the relevant British Standards	British Standards Codes of Practice
Matchboard doors		B.S. 459, Part IV	
Quality of timber and workmanship in joinery		B.S. 1186	
Hinges		B.S. 1227	
Wood door frames and linings		B.S. 1567	
Letter plates		B.S. 2911	
Doors and frames			B.S.C.P. 151
Windows			
Glass — general	Yo		
Curtain walls	(21)		
Ironmongery	Xt7		
Windows			B.S.C.P. 151
Windows — construction elements	(31)		
Linseed-oil putty for wood frames		B.S. 544	
Wood casement windows		B.S. 644, Part I	
Wood double-hung sash windows		B.S. 644, Part II	
Glass for glazing		B.S. 952	
Steel windows for domestic buildings		B.S. 990	
Wood surrounds for metal windows		B.S. 1285	
Steel windows for industrial buildings		B.S. 1787	
Steel windows for agricultural purposes		B.S. 2503	
Modular coordination in building		B.S. 2900	
Recommendations for a system of tolerances for building		B.S. 3626	
Staircases			
Stairs	(24)		
Wood stairs with close strings		B.S. 585	
Partitions			
Plasterwork — general	P		
Gypsum plasterboard	Rh2		
Partitions — general	(22)		
Suspended ceilings	(25)		
Fibre building board for general building purposes		B.S. 1142	
Gypsum building plasters		B.S. 1191	
Hollow-glass blocks		B.S. 1207	
Gypsum plasterboard		B.S. 1230	
Metal lathing for plastering		B.S. 1369	
Precast concrete blocks		B.S. 2028	

FURTHER SOURCES OF INFORMATION · 219

	CI/SfB filing reference	A selection of the relevant British Standards	British Standards Codes of Practice
Blockboard and laminboard		B.S. 3444	
Information about plywood		B.S. 3493	
Asbestos wallboards		B.S. 3536	
Information about blockboard and laminboard		B.S. 3583	
Internal plastering			B.S.C.P. 211
Finishes			
Painting work – general	(D6)		
Paints – types general	V		
Water paints and distempers for interior use		B.S. 1053	
Wallpapers		B.S. 1248	
Classification of wood preservatives		B.S. 1282	
Paint colours for building purposes		B.S. 4800	
Painting of buildings			B.S.C.P. 231
Building Maintenance			
Maintenance – general	(W5)		
Building Alterations and Repairs			
Survey tapes		B.S. 4484	

Index

All references are to pages of text to which the reader is referred for diagrams.

Acoustic clip, 171
Angle of creep, 108
Arches, 52 *et seq.*
 camber, 52
 segmental, 52
 setting out, 52
 skewback, 53
 springing, 53
 template, 53
 voussoirs, 53
Architraves, 135
Ashlar, 22
Asphalt
 fine cold, 13
 hot rolled, 13
Auger, 37
Autoset level, 3

Balustrade, 160
Bat, 49
Batten roll, 101
Bitumen macadam, 9
Bituminous felt, 56
Bituminous paint, 176
Blinding, 39
Boarding – Tongued and Grooved, 90
Bolection mould, 128
Boundary wall, 20 *et seq.*
 brick, 20
 stone, 20
Brick boundary wall, 20
Brick edging, 12
Brick paving, 15
Brick threshold, 55
Bricks, 42 *et seq.*
 classification of, 43
 common, 43
 drying of, 44
 engineering, 43
 facing, 43
 laying of, 47
 manufacture of, 43
 pressed, 44
 rustic, 43
 sand faced, 43
 sizes of, 44
 tunnel kiln, 44
 wire cut, 43
Brickwork, 48 *et seq.*
 bonding, 48
 flues, 61
 flush joint, 47
 jointing, 47
 keyed joint, 47
 load bearing, 26
 pointing, 47
 recessed joint, 47
 weathered joint, 47
British Standards, 212
Builder's square, 4
Building alterations and repairs, 193 *et seq.*
Building maintenance, 185 *et seq.*
Building measurement, 205 *et seq.*
Building paper, 36
Building Regulations, 1, 5, 28, 29, 35, 50, 81, 89, 90, 95, 154
Building Research Digests, 213
Butt hinge, 130

Calcium chloride, 40
Carriageway, 8
Cast stone coping, 55
Cavity walling, 26, 50, 58
 wall ties, 58
Ceilings, 170
Ceiling joist, 118
Cellar fungus, 195

INDEX · 221

Cement, 39 *et seq*
  aggregate, 39
  high-alumina, 39
  mortar, 45
  rapid hardening, 39
  rendering, 57
  sulphate resistant, 39
Chain link fencing, 17
Chain survey, 1
Changes of level, 22
Characteristics of sub-soils, 31
Chlorinated rubber paint, 176
Choice of site, 1
Classification of bricks, 43
Classification of information, 214
Clay floor tiles, 186
Clay sub-soil, 6
Clayware copings, 24
Clear finish, 175
Cleft chestnut fencing, 19
Close boarded fencing, 19
Coarse aggregate, 39
Cobbles, 15
Codes of Practice, 213
Cohesive soil, 32
Cold cured paint, 174
Common bricks, 43
Common furniture beetle, 196
Compound sheet roofing, 100
Concrete
  block walling, 22
  board finish, 86
  curing, 42
  durability, 40
  flat roof, 102
  fencing, 17
  floors, 85
  hearth, 64
  in situ, 14, 86
  kerbs, 16
  mixing, 40, 41
  paving slab, 14
  pre-stressed, 87
  ready mixed, 41
  site, 89
  strength of, 41
  sulphate attack, 42
  workability, 40
Concrete beams
  pre-cast, 87
  pre-stressed, 87
Concrete-block partition, 163
Concrete foundations
  deep strip, 36
  pile, 37
  raft, 38
  wide strip, 35
Concrete frame, 74
Concrete in foundations, 39
Concrete lintel, 52
Concrete walling, 74
Coniophora cerebella, 195
Condensation, 93, 95
Copings
  brick on edge, 55
  cast stone, 55
  clayware, 24
  metal, 25
  pre-cast concrete, 25
  stone, 24
Co-polymer emulsion, 174
Copper flashings, 106, 124
Copper roofing, 99
Copper sheet d.p.c., 56
Coursed rubble, 22
Cover flashing, 107
Cradling piece, 84
Cramps, 72
Creosote, 177
Cross walling, 50
Curtain walling, 26, 148
Cutting and pinning, 26

Damp proof course, 56 *et seq.*
  bituminous felt, 56
  chimney, 126
  copper sheet, 56
  engineering brick, 57
  lead sheet, 56
  mastic asphalt, 56
  pitch polymer, 57
  polythene sheet, 56
  slate, 56
Damp proof membrane, 57, 93
Dampness in building, 193 *et seq.*
Datum, 4
Dead loads, 27, 31, 81
Dead shoring, 197
Death watch beetle, 196
Decibel, 29, 132
Deep strip concrete foundations, 36
Defects in timber, 78
Defects Liability Period, 185
Deflection, 80
Depth of foundations, 31
Distemper, 174
Dog irons, 197

Doors, 127 *et seq.*
  bolection mould, 128
  flush, 130
  frames, 132, 133
  furniture, 136
  glass in, 130
  ledged and braced, 128, 129
  linings, 132, 133
  panel, 128, 129
  sound insulation of, 132
Double lap tiles, 110
Double glazing, 146
Down draught, 192
Drainage excavations, 5
Dry rot, 78, 194
Drying of bricks, 44
Drying out, 204
Dumper, 5
Durability of concrete, 40

Earth moving machines, 5
'Egg Box' partition, 165
Egg shell finish, 178
Electro-osmotic d.p.c., 204
Engineering brick d.p.c., 56
Engineering bricks, 43
English bond, 48
Epoxy resin, 174, 179
Estate road, 8
Excavation, 5
Excavations — safety in, 6
External works, 8 *et seq.*

Facing bricks, 43
Fencing
  chain link, 17
  cleft chestnut paling, 19
  close boarded, 19
  concrete, 17
  interwoven, 19
  palisade, 19
  stranded wire, 17
  timber, 18
  wattle hurdle, 19
  wrought iron, 17
Fill, 38
Fine cold asphalt, 13
Finishes, 172 *et seq.*
Fireclay flueliner, 65
Fireplaces, 61
Fire resistance, 30, 95, 131, 148, 169
Firring pieces, 104
Flashings
  copper, 124
  lead, 124
  zinc, 124
Flat finish, 178
Flat roofs, 96, 102, 104, 106
Flemish bond, 48
Flexible pavement, 9
Floating coat, 169
Float glass, 145
Floors, 77 *et seq.*
  boarding, 90
  concrete, 85
  reinforced concrete, 85
  strip, 91
  tiles, 92
  wood block, 92
Flues, 61 *et seq.*
  blocks, 66
  gas fire, 66
  liners, 65
Flush doors, 130
Flying shore, 201
Footpath, 8
Footway, 8, 9
Formwork, 85
Foundations, 30 *et seq.*
  concrete strip, 35, 36
  depth of, 31
  pile, 31
  raft, 31, 38
  reinforced concrete, 38
  types of, 31
Framed construction, 68
Framing
  fireplace, 82
  timber, 74
Frog, 47
Frost action, 40
Frost attack, 20
Furniture beetle, 196

Gate posts, 20
Gates, 20
Gauged mortar, 46
Glass block partition, 164
Glazed doors, 130
Glazing
  bars, 130
  double, 146
  wall, 147
  wired glass, 145
Gloss finish, 173
Government Publications, 213
Granite setts, 15
Gravel, 14

Ground beam, 37
Ground floor — solid, 92
Gully
  pot, 10
  road, 10
Gypsum plaster, 169

Handrail, 153
Hard core, 92
Header, 48
Heart wood, 77
Hearth
  concrete, 64
  tile, 65
  trimmer, 64, 83
  upper floor, 64
Heat loss, 28
Herringbone strutting, 84
Hot rolled asphalt, 13
House longhorn beetle, 196
Hydration, 36
Hydraulic digger, 5

Industrialized building, 150
In situ concrete, 14, 86
Insulation, 27 *et seq.*
  roof, 95, 103
  sound, 29
  thermal, 27, 148
Interlocking tile, 110
Interwoven fencing, 19
Inverted roof, 102

Kerbs — concrete, 16

Laminated truss, 124
Land drains, 5
Latches, 136
Lattice beam, 87
Laying of bricks, 47
Layout of staircase, 152
Lead
  flashings, 107, 124
  flat roof, 106
  roofing, 99
  sheet d.p.c., 56
  soakers, 122
  tack, 107
Ledged and braced doors, 128, 129
Levelling, 1
Lime cement mortar, 46
Lime plaster, 169
Lining paper, 94
Lintels, 163
  reinforced concrete, 52, 59

Loads
  dead, 27, 31
  superimposed, 27, 31
Load bearing brickwork, 26
Load bearing strata, 32
Locks, 136

Made up ground, 38
Maintenance reports, 187 *et seq.*
Manufacture of bricks, 43
Mastic, 143, 150
Mastic asphalt d.p.c., 56
Matrix, 39
Measured sketches, 207
Measuring rules, 207
Measuring tape, 206
Merulius lacrymans, 194
Metal copings, 25
Metal lathing, 170
Metal sheet roof covering, 99
Mineral surfaced felt, 106
Mixing of concrete, 40, 41
Modular units, 150
Monolithic construction, 26
Monolithic pavement, 9
Moisture barrier, 74
Moisture content of timber, 77
Mortar, 45 *et seq.*
  cement, 46
  droppings, 59
  gauged, 46
  lime cement, 46
  mix, 46
  plasticiser, 46
Multi-coloured paint, 176
Munsell system, 182
Muntin, 129

Newel, 154
Non-cohesive soil, 32

Oil based paints, 173
Open tread staircase, 160

Paint, 172 *et seq.*
  alkyd resin, 173, 179
  application, 179
  bituminous, 176
  chlorinated rubber, 176
  cold cured, 174
  distemper, 174
  emulsion, 174
  multi-coloured, 176
  oil based, 173

Paint *Contd.*
  primer, 178
  roller, 180
  sprayer, 180
  Thixotropic, 176
  water thinned, 174
Painting defects, 181
Palisade, 19
Panel doors, 128, 129
Parapet wall, 58, 103
Partitions, 162 *et seq.*
  clay block, 162
  concrete block, 163
  'Egg box', 165
  glass brick, 164
  plasterboard, 166
  timber, 165
  wood wool, 163
Party walls, 29
Pavement, 8
  flexible, 8
  monolithic, 9
Paving – brick, 15
Paving slab – concrete, 14
Permissible stress, 69
Pile foundations, 31
Pin jointed frame, 70
'Pinning up', 201
Pitched roofs, 108 *et seq.*
  covering, 108
Pivot windows, 147
Plain tiles, 108
Plaster
  cracks, 185
  Gypsum, 169
  lime, 169,
  pre-mixed, 169
Plasterboard, 170
Plasterboard partition, 166
Plastering, 168
Plastic roofing, 192
Plastic wall covering, 182
Plasticised mortar, 46
Plywood, 167
Ponding, 101
Pointing, 47
Polythene sheet d.p.c., 56
Polyurethane, 174, 179
Portland cement, 39
Post and rail, 19
Powder post beetle, 196
Pre-cast concrete beam, 87
Pre-cast concrete copings, 25
Pre-cast paving slabs, 14
Pre-mixed plaster, 169

Pre-stressed concrete, 87
Preservative coverings, 177
Pressed bricks, 44
Primer paint, 178
Profile boards, 4
Promenade roofing, 98
Protected timber, 196
Purlin, 114
Purlin roof system, 114
P.V.C. tiles, 92

Queen closer, 49

Rafter
  function of, 113
  hip, 118
  jack, 118
  trussed, 115, 120
  valley, 118
Raft foundations, 31, 38
Rainwater
  disposal, 12
  outlet, 103
  pipes, 102
  shoe, 12
Raking shore, 200
Rapid hardening cement, 39
Ready mixed concrete, 41
Reasons for painting, 172
Re-plastering, 204
Re-pointing, 191
Reinforced concrete floor, 85
Reinforced concrete foundations, 38
Removal of top soil, 5
Rendering – cement, 57
Ridge board, 118
Ridge tile, 120
Rigid frame, 70
Rising damp, 193
Roads, 8 *et seq.*
  channel, 10
  drainage, 10
  estate, 8
  gradient, 9
  gully, 10
  service, 8
  sub-grade, 9
Rock, foundations on, 32
Roofs, 94 *et seq.*
  falls, 101, 103
  fire resistance, 95
  flat, 96
  insulation, 95, 103
  pitched, 108

INDEX · 225

promenade, 98
screed, 102
snow loading, 96
tile gauge, 110
timber flat, 104
wind loading, 96
Roof covering
  asphalt, 97
  built-up felt, 97, 106
  compound sheet, 100
  copper, 99
  lead, 99
  metal sheet, 99
  thermoplastic sheet, 100
  zinc, 100
Rubber bitumen emulsion, 190
Rubble walling, 67
Running measurements, 208
Rustic bricks, 43

Safe bearing capacity, 33
Safety in excavations, 6
Sand faced bricks, 43
Sand for mortar, 45
Sap wood, 77
'Saw cut' d.p.c., 203
Schedule of Dilapidations, 187
Screed, 92, 102
Seasoning of timber, 78
Segmental arch, 52
Service road, 8
Setting out, 3
Settlement, 31
SfB classification, 214
Sheet floor coverings, 186
Shifting sand, 32
Shop welding, 71
Shoring, 197
Shrinkable clay, 36, 37
Shrinkage of timber, 78, 127
Shuttering, 26, 86
Silicone, 190
Sills
  projection, 54
  stone, 54
  stooling, 54
  throating, 54
  weathering, 54
Single lap tile, 110
Single roof system, 113
Site
  choice of, 1
  drainage, 5
  investigation, 32
  survey, 1

Sizes of bricks, 44
Sketch pad, 206
Skim coat, 170
Skirting, 166
Slate d.p.c., 56
Slates, 111
  battens, 112
  centre nailed, 111
  head nailed, 111
Sleeper wall, 88
Smoky chimneys, 192
Snow loading, 96
Soakaways, 11
Soakers, 122
Soft wood, 78
Solid fuel appliances, 61
Solid ground floor, 92
Solid walling, 50
Sound insulation, 29, 132
Spoil, 5
Spot levels, 1
Staircases, 152 et seq.
  balustrade, 158
  carriage, 155
  dancers, 153
  design points, 153
  dog leg, 152
  fliers, 153
  flight, 152
  geometrical, 152
  layout, 152
  open tread, 160
  open well, 152
  quarter turn, 152
  riser, 153, 157
  stone, 158
  strings, 153
  tread, 153
  winders, 153
  wood, 154
Standing seam, 101
Steel
  structural, 69
  universal sections, 69
Steel tape, 206
Steelwork — structural, 68
Stone
  boundary wall, 21
  copings, 24,
  facing, 66
  masonry, 66
  staircase, 158
  threshold, 55
Stopped end, 48
Straining post, 17

226 · INDEX

Stranded wire fencing, 17
Strength of concrete, 41
Stress grading – timber, 81
Stretcher, 48
Strip flooring, 91
Strip foundations – concrete mix, 40
Structural stability, 26
Structural steelwork, 68
Structure, roofs, 94
Strutting
  herringbone, 84
  solid, 84
Sub-soil
  chalk, 33
  drainage, 23
  peat, 33
  seasonal movement, 32
  shrinkable clay, 36, 37
Sulphate attack of concrete, 42
Sulphate resistant cement, 39
Superimposed loads, 27, 31
System building, 150

Tarmacadam, 13
Template, 53
Terrazzo, 92, 160
Thermal insulation, 27, 148
Thermal movement – glazing, 150
Thermal transmittance coefficient, 28, 146
Thermoplastic
  sheet roofing, 100
  tiles, 186
Thixotropic paint, 176
Threshold, 55, 135
Throat, 140
Tiles
  double lap, 110
  floor, 92
  hanging, 75
  hearth, 65
  interlocking, 110
  plain, 108
  P.V.C., 92
  ridge, 120
  single lap, 110
  under eaves, 122
  Vinyl, 92
Tilting fillet, 112
Timber
  classification of, 127
  fencing, 18
  flat roof, 104
  framing, 74

  ground floor, 88
  partitions, 165
  protection, 74
  shrinkage, 78, 127
  stress grading, 81
  upper floors, 80
Timbering to trenches, 6
Tractor shovel, 5
Trench prop, 7
Trench support – steel sheet, 7
Trial holes, 32
Trimmed joist, 83
Tuck pointing, 47
Tunnel kiln, 44

U.D.C. classification, 214
Universal sections – steel, 69
'U' value, 28
Under eaves tiles, 122
Under floor ventilation, 90
Underpinning, 201 *et seq.*
'Upside-down' roof, 102

Vapour barrier, 74, 95
Varnish, 175
Ventilated under lay, 97
Ventilation under floor, 90
Verge, 123
Vinyl tiles, 92

Waling, 7
Walling, 26
  cross, 50
  solid, 50
Wall electrode, 204
Wallpaper, 181
Walls
  cavity wall, 58
  curtain walls, 68, 148
  fire resistance, 30
  glazing, 147
  panel, 68
  sleeper, 88
  sound reduction, 29
Water bar, 135
Water–cement ratio, 40
Water-thinned paint, 174
Water for concrete, 40
Water for mortar, 45
Water table, 32
Weather boarding, 75
Weigh batching, 41
Welding – shop, 71
Welted seam, 101

Wet rot, 195
Wide strip concrete foundations, 35
Wind loading — roofs, 96
Windows, 138 *et seq.*
   aluminium, 126, 142
   analysis, 139
   casement, 140
   metal, 141
   module 100, 141
   opening light, 140
   pivot, 147
   principles of framing, 138
   sliding sash, 144
   standard, 141
   standard agricultural, 142
   standard industrial, 142
   weather stripped, 141
Wire cut bricks, 43

Wood
   block floor, 92
   blocks, 186
   dry rot, 78
   moisture content, 77
   sap, 77
   seasoning, 78
   shrinkage, 78
   worm, 196
Wrought iron fencing, 20

Zinc
   chromate primer, 177
   flashing, 124
   oxychloride plaster, 195
   roofing, 100